WITHDRAWN BY THE
UNIVERSITY OF MICHIGAN

D1526221

# TECHNOLOGY AND RATIONALITY

# Technology and Rationality

THOMAS KROGH

**Ashgate**
Aldershot • Brookfield USA • Singapore • Sydney

© Thomas Krogh 1998

All rights reserved. No part of this publication may be reproduced, stored in a retrieval system, or transmitted in any form or by any means, electronic, mechanical, photocopying, recording or otherwise without the prior permission of the publisher.

Published by
Ashgate Publishing Ltd
Gower House
Croft Road
Aldershot
Hants GU11 3HR
England

Ashgate Publishing Company
Old Post Road
Brookfield
Vermont 05036
USA

**British Library Cataloguing in Publication Data**
Krogh, Thomas
    Technology and rationality. - (Avebury series in philosophy)
    1.Technology - Philosophy 2.Technology and civilization
    I. Title
    601

**Library of Congress Catalog Card Number:** 97-77556

ISBN 1 84014 103 4

Printed and bound by Athenaeum Press, Ltd.,
Gateshead, Tyne & Wear.

# Contents

| | |
|---|---|
| *Foreword* | vii |
| Introduction | 1 |
| **PART I: ANTHROPOLOGICAL THEORIES OF TECHNOLOGY** | 15 |
|     Chapter 1    Ernst Kapp | 17 |
|     Chapter 2    Arnold Gehlen | 25 |
| **PART II: THE DIALECTICAL TRADITION** | 37 |
|     Introduction to Part II | 39 |
|     Chapter 3    Georg Wilhelm Friedrich Hegel | 41 |
|     Chapter 4    Karl Marx | 57 |
|     Chapter 5    Jean-Paul Sartre | 81 |
| **PART III: THE COMMUNICATIVE THEORY** | 97 |
|     Introduction to Part III | 99 |
|     Chapter 6    The Theory of the Technical Cognitive Interest | 103 |
|     Chapter 7    Technology as Ideology and a Form of Action | 121 |
|     Chapter 8    The Critique of the Philosophy of Consciousness | 143 |
|     Chapter 9    Technology between the Lifeworld and the Systems | 153 |
|     Chapter 10   Technology in the Lifeworld | 171 |
| *Bibliography* | 195 |
| *Index* | 203 |

# Foreword

Most of this book was written in the period 1990 to 1992 when I had a three-year research grant from the Programme for Cultural Research, part of the former Norwegian Research Council for Science and the Humanities (NAVF). I would like to take this opportunity to thank NAVF for the grant which enabled me to study theories of technology, including a stay at the Freie Universität in Berlin. The University of Oslo, Department of Cultural Studies, provided financial support for this publication.

During this period I worked in close contact with the fledgling Centre for Technology and Culture at the University of Oslo. This was a dream come true for me as I have always wanted to work in an interdisciplinary environment consisting of philosophers (or at least one!), historians and social scientists. Any interest in this book for people outside the perhaps rather insular world of professional philosophers and historians of ideas is due to the environment at the Centre during the period when I was there. The structure of the Centre was such that there was always a large number of researchers with very varied backgrounds working there. This profile, which is a major strength of the Centre, unfortunately makes it impossible for me to thank everybody individually for their help and support.

There are however two colleagues to whom I feel particularly indebted: Francis Sejersted and Sissel Myklebust. The reader will soon discover how much I have learned from Dag Østerberg and his pioneering work on the concept of materiality and social materiality. Gunnar Skirbekk read an early version of this book and gave me some very constructive criticism and ideas for improvements, especially in connection with my discussion of Habermas. I have also benefited greatly from my on-going on-line discussions with Andrew Feenberg.

Thanks are also due to Jens Erik Paulsen for the layout, and to Sissel Myklebust and Rigmor K. Johnsen for the work of coordinating and finalizing the manuscript for publication in August 1997. Finally I must express my deepfelt gratitude to the translator, Jenny Gillott, for her very professional and thorough work with a complex Norwegian manuscript.

All errors and weaknesses in this book are of course my own.

Oslo, August 1997
Thomas Krogh

# Introduction

Technology studies seem to be haunted by an eternal dilemma: On the one hand, we have an ever growing awareness of the fact that technology is a fundamental factor in all areas of human life and activity, and on the other a lingering insecurity about how the phenomenon of technology can be brought into focus; at what level of analysis and with which theoretical tools we best can dissect it.

There are many approaches to this problem today, ranging from a purely engineering approach, which ends up simply describing individual technological objects, such as machines and apparatuses, at one end of the scale, to a historical approach to the modern sociology of technology and science, which attempts to integrate technology entirely into the political and cultural processes, at the other. Both of these extremes are equally unproductive, as they do not deal with what I deem to be the most fundamental hermeneutic preconception in the arena of technology studies: That technology never exists on its own and, in general, ought never to be studied separately from social, economic, political and cultural processes, and that technological processes at the same time have an internal dynamic which does not coincide with any of these other forms of human activity.

This contention may appear to be rather weak, as indeed it is meant to be, since its function is quite limited. It expresses what the American philosopher, Felix Ferré, called the omnipresent function of technology (Ferré 1988, p. v). In a different context I have myself used the formulation that technology is the main reality-producing factor of our era. This temporary clarification is not however adequate to disentangle the disorganized mess of the mutually differing and irreconcilable, rudimentary theories and approaches.

It is tempting to assume that this is precisely due to the significance of technology to *all* forms of societal activity; that it is the impact and effect of technology on the widest range of fields of human activity that have created this chaotic picture. Technology is not a research object that can easily be fitted into the traditional scientific divisions of labour. There is probably no one satisfactory solution to this problem that is able to provide us with the ultimate method for technology studies or lay the foundations for a new superdiscipline. We need interdisciplinary cooperation between the various autonomous sciences with their established methods. One ought, then, to be sceptical right from the start to the term philosophy of technology. Studies of technology or philosophy of technology which can be clearly recognized as philosophical must always be capable of entering into this kind of interdisciplinary interaction with empirical disciplines, and will often have been inspired by them, even if they are not capable of contributing directly to these studies.

This book cannot of course cover the whole field that I have indicated here. It builds on a selective starting point, and deals with the position and role that the technology of modern society has traditionally had and ought to have within modern social philosophy and social theory. As will become clear in the following chapters, these two considerations are by no means always one and the same. This selectivity is expressed on two levels: My choice of historical writers who are considered; and in the systematic orientation towards a discussion of the way in which technology is treated in social philosophy or social theories that have been inspired by and reflect philosophy, and that move on a relatively high level of abstraction, such as those proposed by Gehlen, Marx, Sartre, Habermas and Luhmann.

Herein lies the very reason, however, that there may be grounds for trying to delimit the area of the phenomenon we call technology and for narrowing it down in relation to other areas of society. In this case it is undoubtedly tempting to start with the fundamental idea that the role of technology lies in the processing of external nature, and that technology must therefore first be defined in relation to the classic dichotomy between society and nature.

It is this intuition that Ferré appears to have let himself be steered by. He introduces three concepts of nature (Ferré 1988, p. 28) which may be summarized as follows:

Nature 1: Everything but the artificial
Nature 2: Whatever happens or exists in time and space
Nature 3: Whatever happens on the basis of immanent processes.

In my opinion there is another division which can advantageously be laid on top of and across these distinctions, namely that between human and non-human nature.

If we now imagine an area which is Nature 2 minus Nature 1, we end up with the area for technology, the entire sum of artefacts, including both those artefacts which are used to produce other artefacts and these produced artefacts. The distinction between means of production and product does not mark a line between technology and its results, but rather lies within the sphere of technology itself.

Ferré's scheme relies on fundamental intuitions found in the best-known discussions of technology in the recent history of ideas, e.g. in Marx and Sartre. Society is perceived as an aggregate of subjects who, by processing nature, control, secure and reproduce their conditions of life. We have defined an area for technology which now appears as a regional ontology in the phenomenological tradition. This school was without a doubt also Ferré's background.

The first problems we encounter with his scheme arise when we ask whether it allows us to draw precise boundaries between the various fields. In a tradition that is coloured more by dialectics than phenomenology, and that is also supported by Marx and Sartre, one tends to emphasize the fact that labour and technology imply a mutual mediation of society and nature. The world, as it appears to us, is basically characterized by human activity, at the same time as technology is both subject to the laws of nature and itself presupposes them in order to be able to function. Viewed from this angle it becomes meaningless to ask where nature begins and society ends. (See in particular Marx 1969a, p. 41f. and Schmidt 1963, passim.)

Nevertheless, the observation does not undermine Ferré's argument: We are still talking about two fields of reality and the forms of their connection. We take the decisive step when we start giving up hope of finding an ontologizing approach, like the one Ferré suggests. The point is not so much that modern technology has such global consequences that it is no longer possible to find any geographical region untouched by human intervention. It is more that even a machine is still a part of nature, in the sense that its operations are regulated by universal laws. The same is true of a cultural landscape which has been developed over millennia.

The distinction between Nature 1 and (Nature 2 minus Nature 1) should therefore be seen not as an ontological distinction, but as a result of the perspectives we at any one time have chosen to adopt towards the objects we are dealing with. A machine as well as a cultural landscape can thus be viewed from two different angles, depending on which perspectives one prefers to use as a starting point: They can be seen both as an object of study for the natural sciences (and in this sense there is a "pure" nature) and as objects which have appeared through intentional human activity connected to machines and tools.

A somewhat audacious summary of this could be: Of course the use of any form of technology requires insight into the laws of nature, as these appear to the users from the insight into nature that the individual technology requires. This does not however mean that it is most advantageous to view technology as a relationship between society and nature, or to make a basic distinction between society and technology. Technology can be regarded most fruitfully, but not exclusively, as a part of the historical development of human society. The concepts we apply to artefacts and the descriptions of them which make use of these concepts, must be taken from — or at least be applicable to — a general theory of society, not from natural science or engineering science.

This is emphasized clearly by the German historian of ideas, Hans Blumenberg, in one of his earlier articles:

> Neither the antithesis between nature and technique, where nature appears as the difference when the layer of culture has been subtracted, nor taking a natural technical ability as our starting point, bring us to the core of the problem that technization is a process which appears suddenly in history. This process no longer appears to have any reasonable explanation in human nature. On the contrary its imperatives of adaptation become unavoidable in contrast to a nature which now appears as deficient (Blumenberg 1963, p.9).

It is at any rate this last perspective which must form the basis for sociological and philosophical theories of technology, regardless of how much more useful it may be from an engineering point of view to make use of a more limited angle. The decisive point here is that the concept of an object does not belong only and exclusively to natural science and traditional epistemology, which took as its starting point the concept of an object used in the natural sciences. There is also a tradition of operating with a social concept of the object or the thing, on a par with the concepts of people and actions, a tradition which first saw the light of day in connection with Roman law. (See Gratian *Digest* I, 5,3., quoted from Donald Kelley in "Civil Science in the Renaissance" in Pagden 1987, p. 64.) In the triad *personalitas, realitas, actio*, according to Kelley, we find a legal language that can describe and normatively regulate a social reality and which includes concepts for both persons and things and sees them as connected through the concept of acts of will (Kelley in Pagden 1987, p. 64).

In this and similar traditions the thing is not first and foremost a part of the objectifying scientific understanding in the natural sciences, but rather an object and mediator of needs, desires and communicative connections. In this way the object and the artefact appear as an object for the social sciences' way of understanding research objects.

What Kelley says of things can equally well be applied to all artificially created objects:

> In particular Gadamer has pointed out that the category of 'thing' as well as that of 'person' must, to be meaningful, be interpreted in a social sense. In other words *res* must be understood not according to the old Cartesian duality (*res extensa, res cogitans*) or the Kantian *Ding an sich*, but rather as the res of civil law, which represents an extension as

well as an object of consciousness, desire and sociability (or belligerence) — which is to say of personality, action and politics (p. 77).

But in exactly the same way as action, systems can reach a complexity where what takes place on the macro level goes a long way beyond, and can indeed even become the opposite of, what was intended by the individual actor, so can also the systems of social things that constitute modern technology assume a character of contrafinality, of an apparent internal dynamic (and objective exigencies) that crosses the plans of the actor. To put it simply we could say that, at any rate in modern societies, we are confronted with a technology in which things bear the characteristics of institutions.

The problem for so much of technology theory, and particularly the politics of technology, seems to be this: In the field of technologies *inputs* do not consist of anything more than human intentionality in the form of decisions, prioritizations and interests. What comes out seems to be something else and something more, as if technology had an immanent logic independent of all other relationships, not least of our intentions. In this way technology seems to have an especially close connection to what the dialectical tradition has described as alienation and reification; not so much in that technologies consist of things, but that they assume the characteristics of things, such as putting up resistance, of not necessarily evaporating in our actions. This may seem somewhat enigmatic, but it is no more enigmatic than, and not enigmatic in a different way from, the complexity which permeates almost all relationships in our developed modern society. This enigmatic quality appears inevitably also in connection with technologies when they assume an institutional character.

Incidentally, we can relate the theory that seems most likely to pave the way for this kind of understanding to the notion of *materiality* as developed in the phenomenological tradition. Its roots stretch back to ideas found in Hegel and it was developed in Marx's materialist conception of history. The theory has been developed further in our era by Sartre, and in Norway by Dag Østerberg in particular. (For further references see Part II and Part III, chap. 8.)

What we call materiality can be defined as a social reality of objects produced by human activity. We must suppose that this productive activity on the micro level is a conscious activity that must be interpreted in the light of the needs, plans and insight of the actors. Of course we must be careful not to assume that the consequences of the individual acts are controlled and understood. This caveat concerns individual acts studied on their own and the aggregated results. Materiality will then consist of 1) the means of the technological processes (instruments and tools), 2) the results of these processes (the products), and 3) the social reality generated through a technological process. The factory system from the end of the eighteenth century is a phenomenon which is material in senses 1 and 3. A machine in a factory and a word processing network are forms of materiality in senses 1 and 2. Both urban landscapes and cultural landscapes are forms of materiality in sense 3.

It is, then, my opinion that the concept of materiality offers the best starting point for considering technology in its social context in modern society. This point of view has been summarized excellently by Bernward Joerges:

> Above all in Marx....[there are] similarities to a concept of technology according to which machines not only carry out the metabolism between nature and society, but at the same time are nature and society. To talk about the increasingly *social character of nature* does not only imply that it becomes useful in technical ways through natural or engineering sciences, that it becomes atomized and transformed into products and means of production for various purposes. But in this way, from a sociological point of view, things acquire a character of quite specific institutions and structures of competence (Joerges 1988, p. 28).

I would like to stress again that this view of technology has no ontological pretensions: I am consciously trying to avoid an ontologicizing point of view. I am recommending this starting point and the concept of materiality that follows from it not because I think it deeper or truer than the objectivating approach which forms the basis for the concept of nature in the natural sciences, where nature is quite correctly perceived as separate and independent from society. From an ontological point of view it is probably equally "correct" to regard a radio from a physical, economic or social point of view, that is to say, to regard it as an application of certain physical principles, as a result of certain investment strategies, or as a link in the distribution of power and impotence in society. The advantage of taking the concept of materiality as a starting point is quite simply that it provides a way of understanding the forces which form and guide technical developments, which the natural sciences and engineering science are unable to accommodate.

This point of departure is different from, but probably compatible with, the attempts we find in modern phenomenology to consider natural science itself as grounded in a prescientific lifeworld, but I regard it as completely separate from these attempts and do not intend to try to develop my analysis in that direction.

In this introduction I will touch upon two questions which have made their presence very much felt in the general debate about technology and within the theory of technology in recent times. These observations, which probably seem rather sketchy, can serve as a further presentation of the perspective I have been hinting at.

The first question which it seems natural to bring up in connection with this kind of approach, is whether there is any point in talking about various forms of technology as natural and unnatural. Needless to say, there is more than one correct use of the terms natural and unnatural in this context. There are several different distinctions that can be used to mark a possible line between natural and unnatural technology. But I would like here to present two different arguments to support the view that this sort of use of language is unfortunate and misleading.

Firstly, I would simply point out that there are very few examples of "natural" models for technological developments made by man, natural here in the sense of both the so-called external natural processes and man's own nature. The best examples of important inventions which had no model in nature are the controlled use of fire and the wheel. There is no such thing as mimetic technology, or at least technology in general cannot be regarded as mimesis of natural processes or bodily functions (see below, Part I).

My second argument is connected to the historical emergence of technology in the form of the use of simple tools. Technology has been an important moment in natural history since the transition of the various species (their ethology) to the process of civilization. Even primates demonstrate a remarkable ability when it comes to the invention of tools and improvization regarding their use. This could of course be used as an argument in favour of the view that technology is founded in human physiology. The point is however rather that evolution in primates has already reached a level where creativity and improvization are possible. In humans too technological operations must be linked to and remain within the framework of possible, biologically determined, stimulus-response connections, the motor requirements for movement of limbs and for coordination of perception and movement. But the use of technology always goes beyond our forms of behaviour which are ruled by instinct. We must therefore be reluctant to claim that use of even the simplest forms of technology is natural to human beings, as the use of tools by definition breaks out of the bounds of behaviour that can be regarded as biologically determined. (The general possibility for this kind of behaviour must of course be recognized as a result of the evolutionary process, but not the individual characteristics of behaviour which are related to technological development.)

From this we can conclude that a technology can be effective or ineffective, healthy or harmful, dangerous or harmless, radical or conservative, ugly or beautiful, alienating or unifying; in fact about the only thing it cannot be is natural, original or correct. *In relation to* a certain level of cultural development, we can of course call technologies that cause illness or fatigue unnatural, for example many forms of factory work. And many representatives of the ecology movement of the last few decades have testified that they believe that use of premechanical technologies is more natural, more suited to the rhythm and motor mechanisms of our bodies, than use of mechanical technologies. My objective here is not to cast doubt over the validity of these statements as expressions of an individual experience of life, but there can be little doubt that there are also people who find great satisfaction in living in and being integrated in highly technological systems. Competitions of authenticity are however a dubious business, and the point is that the phenomenon which can be called technological alienation must to a great extent be a historically relative phenomenon, perhaps even a symptom of the generation gap.

So either we have to say that all technology is unnatural because it breaks with an instinctive function circuit of stimulus-response reactions, or we can say that all technology is natural because it is anchored in the openness that characterizes human nature (and that of some primates). I lean towards the second statement because it best describes the relationship between instinctually controlled behaviour and the use of tools. But the fact that both statements are plausible must surely lead us rather to the conclusion that the distinction between natural and unnatural forms of technology is basically invalid. (The distinction between the use of renewable and non-renewable sources of energy is however a horse of a completely different colour, and as far as I can see a totally valid distinction. But since even some Neolithic cultures destroyed their own ecological foundation, I do not think that this can be used to distinguish between natural and unnatural forms of technology.)

Another question we ought to be able to put behind us is that of the relationship between society and technology. There is after all no relationship *between* society and technology.

This must of course be stated more precisely. In saying this I do not wish to declare any allegiance to the research programme called "SCOT", "the social construction of technology", which has adopted the metaphor of "the seamless web" to describe the position of technology in society. (For a programmatic presentation of the SCOT programme, see Bijker, Hughes et al., 1989.)

No doubt this research programme has latched on to an important intuition: Technology can no longer be understood as restricted to the sphere of labour. But it does not really get to grips with its omnipresent existence. For one thing, it is and will always be a mistake to dismiss any kind of inner division of society into different fields on the grounds that there always turns out to be some sort of connection between the fields. In one sense the application of technology in heavy industry and the institutionalization of technological knowledge in certain institutions of research and education are of course examples of differentiated societal subsystems. One can also claim that the concept of society with which the SCOT programme operates is inadequate, that the elements of power and economics are conspicuous by their absence.[1] However, this is not my main point.

The point of view for which I am arguing in the following is briefly that technology, understood as materiality, is an aspect of and sustains all social relationships, to varying extents of course. And it is in precisely this role that it receives its momentum that has so often been misconstrued as technological determinism or misinterpreted in those theories that sail under that name. The point then is that technology and "society" are not related as explanandum and explanans, as the constructivists in the SCOT programme claim.

It is convictions of this kind which have determined the course of this book. But first some negative delimitations.

First of all let me make it quite clear that this work does not belong to the school of thought that in various countries and at various times has been called the philosophy of technology. I may have started my account with the book that first introduced the concept, *Philosophie der Technik* by the German philosopher, Ernst Kapp (Kapp 1877), but this was after all a book which, despite its speculative and slightly Epigonic character, drew attention to perspectives which were to prove fruitful much later. The same cannot be said of the subsequent German *Technologiephilosophie* from the turn of the century and onwards. From the point of view of the history of ideas, it can be seen, thematically or as a genre, as an undercurrent within what in Germany is called philosophical anthropology: General speculations about the nature of man and his role in an alleged total cosmic or natural-historical framework. In its tendencies it fitted in pretty neatly with the conservative criticism of culture, the German establishment's generally antiliberal opposition to the basic characteristics of the development of the cultural and political forms of the modern world (Ringer 1969). From the perspective of sociology of professions, the far from insignificant popularity which this sort of pondering over technology enjoyed in Germany can be regarded as the result of the difficulties the German technological intelligentsia experienced in finding a foothold in the academic community (Popitz, Bahrdt, et al. 1957, section I,1). The genre *Technologiephilosophie* and the debate fora in which it developed became a vehicle for arguing that engineering, as a science and

---

[1] For a good critique of this point, see Ingunn Brita Moser's thesis *Teknologi i Samfunnsteori: Forskyvninger og forflytninger (Technology in social theory: Displacements and transfers)*, Moser 1993, chapter 6, especially p. 170 ff.

a profession, could be made to fit into the inherited conservative and romantic conceptions of the relationship between the individual, society and nature, which were such an important element in German ideology.

The disciples of *Technologiephilosophie* can thus be described, to use the expression of the American historian of ideas, Jeffrey Herf, as "reactionary moderns", an ideal type he uses perhaps somewhat indiscriminately (see Herf 1984). This tradition is of little philosophical interest, and must not, as Herf tends to, be confused with the works of professional philosophers such as Max Scheler, Martin Heidegger and Arnold Gehlen. Their contributions are of course on a completely different level, even if the latter two shared some of the unpleasant political characteristics of the philosophers of technology — and had a few more of their own design.

A "philosophy of technology" has also developed in the USA in more recent years. It has made its presence most strongly felt in the journal *Research in Philosophy and Technology*. At its best this tendency considers itself as explicitly interdisciplinary, rather than as a new, internal subdiscipline of philosophy. Despite its relative youth, we can already make out two distinct main trends, in addition to the wing that deals purely with the history of ideas. There is one group that is oriented towards political philosophy. It discusses the development and function of technology, and its significance for economic organizations, political institutions and in particular the democratic ideals of modern society. The best known exponent of this research interest is Langdon Winner (see Winner 1977 and Winner 1986).

The other interest can perhaps be more accurately described as social philosophy, as opposed to political philosophy. Here we find studies of modern, technified everyday life, of the individual's use of and relationship to technological apparatus, and of the kind of knowledge and reactions that are produced by contact with modern technology. It is worth noting how isolated this branch is from the dominant trends in American philosophy of science, both in its positivistic and post-positivistic phases. Its inspiration originates mostly from interwar European philosophy, from phenomenological philosophy and from the anti-naturalist tendencies in West-European Marxism, namely Lukacs and the Frankfurt school. Heidegger in particular moves heavily upon the face of the waters. Don Ihde is perhaps the most prominent proponent of this lifeworld theoretical view of technology (see Ihde 1990 and Ihde 1991).

Nor is this book intended as a work of the "From flint axe to nuclear reactor" type. I will not refer the reader to other works of this kind either, not because they do not exist, but because they are all worthless. One can of course imagine the possibility of writing a comprehensive history of technology and inventions that covers the developments from the Great Rift Valley to Silicon Valley, with room for a discussion of the interplay between technology and culture. (It is thought-provoking that nobody has managed it yet and that we are obviously still an enormously long way from such a goal.) But my point here is that quite simply there is no one single phenomenon that we can label as *the* process of technological development throughout the entire history of mankind. There is no *homunculus faber* hidden in us all. A phenomenon like that is just as much a chimera as Technology and *the* technology. I therefore prefer, wherever it is grammatically possible, to use the expression technology without the definite article. In that way I am not limiting myself terminologically should there turn out to be one single, one dominant, many related or a whole plethora of technologies and technological directions of development in a given period in history.

In addition, my theme here is primarily the role that technology plays in theories of *modern* society. That is why there is no independently developed anthropological or paleo-anthropological perspective in this work. I refer to such theories only when the question arises as to the extent to which superhistorical or completely general theories of technology can contribute to theories on a high level of abstraction that focus on technology in modern societies. The question of the role of technology in what can be called the transition from the ethology of animals to the human process of civilization is of course both fundamental and fascinating, but in this book it will only be dealt with in the context already mentioned.

Nor is this book meant to be an introduction to what we can call the tradition of technology criticism in the history of ideas in modern times. Such theories generally operate on a meta-level with regard to the question of how to discuss technology within the framework of a general theory of society. A responsible discussion of these theories would have to define their positions in relation to the various forms of criticism of culture and social criticism that have made their mark on our society since the Enlightenment. A discussion of this kind of theme, which belongs to the history of ideas, would therefore go beyond the framework of this book and take place on a different level to that of the subject I have chosen to discuss here. That is why writers like Ellul, Heidegger, Mumford and Marcuse are not discussed in this book. For a discussion of the criticism of technology, see Krogh 1991a, Krogh 1991b, Sieferle 1984 and Winner 1977.

This book is divided into three main parts: Part I consists of some of the more general anthropological theories of technology (Kapp, Gehlen), which claim to place technology within the framework of a superhistorical concept of man, or at least a total theory of the history of human development. As I have already mentioned, I am highly sceptical to whether these theories actually hold water.

In Part II I turn to what I call the dialectical tradition (Hegel, Marx, Sartre). One could say that the schools of thought that we know as sociology of technology and history of technology have their foundations in this tradition, even if the continuity has not been completely unbroken.[2] More or less explicitly, all of these writers have their point of departure in the technology which became central to society with the growth of the factory system at the end of the eighteenth century. This is a technology which can only be understood when economic categories are also taken into account; machines were the fixed capital and were of decisive social importance in the spread of the employer-employee relationship. Even if the new classes, which were defined by the relationship between labour and capital, were not in fact brought into existence by the development of machines and factories, by the coal-iron complex (and it is impossible to prove that they would not have developed without them), it was nevertheless through this type of technology that the class system came to form the basis of modern society.

However, this form of materialistic phenomenology is based on a fateful theoretic fallacy or one-sidedness, which cannot be overcome on its own terms. So far I have concentrated on the concept of materiality as an expression of how processed things have a social character and serve both to reproduce and change the social system. But the philosophical basis for this tradition itself remained tied to the classical subject–object model in modern epistemology. Society was still conceived of as a supersubject which was connected to its surroundings, i.e. external nature, according to a model of labour, if not of experience. If we are going to continue to use the concept of

---

[2] This kind of view is found in, for example Rosenberg, N. "Marx as a student" and *Inside the Black Box: Technology and Economics*. Cambridge, 1982. See also Lenk (1982), *Zur Sozialphilosophie der Technik*. The first scientific discussion of technology and the origin of the term itself is found in Beckman 1780 (1970), *Anleitung zur Technologie*. Beckmann combined technological and economic evaluations right from the start. His book, which was intended to be of use to both capitalists and state bureaucrats within the framework of his contemporary German system, commercialism, dealt with subjects like the price and quality of raw materials and the means of productions. In the true spirit of rationalism, which by then was on its last legs, he also tried to provide an overall picture of the nature of the multifarious processes of production, based on their fundamental principles. Despite his early modern economic theory (for example he presupposes that the products will be traded as commodities and emphasises the decisive economic "drive" in the lower classes), he got bogged down in the Aristotelian definitions of craft when it came to the technological description of the production processes. Craft is the processing of already available materials and technology; it is the knowledge of, not in, these processes. He failed in his attempt to provide a similarly systematic description of the structure of professions, choosing to fall back on the traditional divisions of the guilds. For a philological overview concerning the development of terms which were prerunners to technology, from the Greek *techne* to Beckman, see Siebke 1968.

materiality then it will have to be reformulated and adapted to a completely different philosophical system.

The philosopher who has most convincingly demonstrated the faults in the model that gave birth to the concept of materiality is Jürgen Habermas. In Part III, then, I introduce his social theory, which I deem to be the most promising theoretical starting point today, and ask both whether it cannot fruitfully be used to understand the role of technology in modern society, and whether it will not then be possible to overcome the lack of focus on technology in Habermas's own work.

My concentration on Habermas is therefore the distilled result of two separate moments. Firstly, the dialectical or materialist–phenomenological tradition was tied up with what Habermas calls the subject–object theory of society. As I mentioned earlier, I find his critique of this quite convincing, and believe that it is therefore necessary to reformulate the results of the dialectical tradition in a new light. Secondly, the actual historical presuppositions, particularly for the way Marx viewed technology, have become outdated. It is no longer adequate to perceive technology as the technology of production, as connected only to the sphere of labour and its organization. Technology, or technologies, embrace the whole of the social reality today. The theory of basis and superstructure can still be accepted as a theory of the field in which historical change originates, but it is not sufficient as a theory of technology today. I regard this starting point as totally in keeping with modern tendencies within sociology of technology as expressed by Gershuny (Gershuny 1983). But here too we must make a distinction. I argued earlier against the exaggeratedly culturalistic conception of technology of the SCOT programme. I am not primarily interested in technology in the pores of everyday life, but in its role, positive and negative, in the reproduction of the lifeworld itself, beyond and in addition to its material reproduction. It is therefore worthwhile joining Werner Rammert (Rammert 1988) here in filtering out conceptions of the function of technology that go beyond the pure sphere of labour, making "everyday life" a cultural category or an isolated research paradigm, and sticking to Habermas's double perspective on society as *both* lifeworld *and* system (Habermas 1987b vol. II, p. 110 ff).

Habermas's theoretical productions are many and complicated, and even the critique of the subject–object model does not appear until halfway through his philosophical development. In chapter 6 and 7, I look at his early theories of technology and discuss first his view of the relationship between science and instrumental actions, and then his theory of the role of technology as ideology in modern society. In chapter 8 I outline his critique of the subject–object model. In chapters 9 and 10 I will try to prove that technology can be dealt with satisfactorily in his new theory of society from 1981, which is based on the theory of communicative action. I will attempt to show that the theme technology is not as far-removed from Habermas's sphere of interest in his later works as is often claimed. In this book I am aiming to demonstrate that the most suitable starting point for developing a theory of technology within the framework of Habermas's theory is his concepts of the two media, power and money, and his distinction between the two forms of integration, social integration and system integration.

# Part I
ANTHROPOLOGICAL THEORIES OF TECHNOLOGY

# 1 Ernst Kapp

First of all I would like to iterate that in this book I am not using the term "anthropology" primarily in the sense of the academic subject which in Anglo-American (and Norwegian) contexts goes by that name. In this part of the book I will be dealing with two writers, Ernst Kapp and Arnold Gehlen, who are most commonly associated with the movement in Germany known, especially after the turn of the century, as philosophical anthropology. I have to admit that I am somewhat sceptical about the fruitfulness and scope of adopting such an ahistorical approach to technology and technological change. There is certainly good reason to be sceptical of Gehlen's use of his philosophy of technology as a part of a general theory of the history of civilization and of the processes of socialization that mankind goes through.

I will voice my criticisms of these writers along the way, rather than saving them up for a showdown at the end. By way of an introduction, however, it might be useful to highlight the advantages of applying the kind of perspective that these writers use. Theories within the philosophical–anthropological school of thought accentuate the fact that man's primary relationship to his surroundings is one of change and alteration, i.e. that the concept of surroundings must be related primarily to *action* rather than to *observation*. The concept of action which is relevant in this context must be defined more precisely. First of all we must differentiate between, on the one hand, action in the sense of using, applying and producing artefacts, above all tools — what we can call instrumental action — and on the other hand, various kinds of social interactions, both strategic and communicative. The writers I have chosen to study here deal for the most part with action in the first sense. I will return to their views on the relationship between technical actions and social development in my critical discussion.

A different theory with a somewhat more limited scope which has been developed in this anthropological tradition, especially by Arnold Gehlen, deals with the relationship between actions in their role as elements of technical operations, and the structure of scientific theories and experiments. Gehlen claims that the technological actions involved in even primitive use of tools constitute the basis for man's experiments with nature. I am going to pass over this particular problem here, but will return to it later (in chap. 6), where I will deal with it in connection with Habermas's version of this thesis.

There are, however, still grounds for asking ourselves to what extent awareness of technological development can be said to have coloured the general ideas we have about nature in our culture. The kind of theories about technology that we come across in Kapp and Gehlen are well-suited for disclosing man's relationship to his surroundings that must have dominated original human societies, and which in one sense is still a precondition for the continuation of human societies' reproduction of themselves.

Nevertheless, regardless of the validity of the theory of the fundamental significance of technology for the basic structure of experimental physics, there is still good reason to doubt that theories of nature, which originated in Greek philosophy, generally had their starting point in or have been characterized by a reflection over man's active relationship to his surroundings. Theories of a contemplative nature have, at any rate in certain eras, characterized the development of both general interpretations of nature and scientific theories in a crucial way.

There are probably several different routes that lead to an interpretation where the concept of nature at a theoretical level can also be linked to ideas about technology and technological advances, i.e. that nature is connected to what can be *done*. But this point is definitely reached when the concept of knowledge is not only associated with what nature is, but also with what can be done to it and with it. There is, however, no evidence of this kind of conception to any noteworthy extent before the seventeenth century. It is common today to read these kind of views into Bacon, as well as Descartes. Kurt Lenk finds a similar concept of technology-compatible nature in Descartes in the fact that: "Nature is no more the objectively given, but the objectively possible". In connection with Bacon he uses the expression "the concept of possible nature", which he has borrowed from Wolfgang Krohn.

> This concept of nature transcends the simply contemplative and conceptual way of regarding nature. For the first time the operative consideration of nature becomes linked with the conceptual one. (Lenk, 1982, p. 254, see also the other literature quoted there.)

There is a parallel in Hans Blumenberg's work regarding the point that Lenk underlines in connection with the concepts of knowledge and nature. Blumenberg claims that the German seventeenth-century metaphysicist Christian Wolff's definition of matter as "a kind of limit to action" (*actionis quasi limes*) paves the way for a worldview which allows human activity the widest possible scope (Blumenberg 1973, p. 256).

I am not going to pursue these constructions in the direction of the idea put forward by Heidegger and his pupils that all modern philosophy is an expression of technology, and thus an expression of a need for control and mastery. (For a typical example see Baruzzi 1973). I would merely suggest that if interpretations of nature as an arena for action, rather than as a given order, are taken to be signs of a technological attitude to reality, then this must be a result of the fact that technological development in the seventeenth century for the first time was reflexively available, experienced as a characteristic of the historical development (see Blumenberg 1973a, Part II, ii). Something must have happened within the development of technology, something new that made it visible for the first time, made it open to reflection, as I expressed above. This new attitude cannot therefore itself be a direct expression of the general, ahistorical constants that an anthropological theory of technology aims to reveal, a fact that in Blumenberg's opinion is an argument against Gehlen's theories. I assume that he had Gehlen in mind when he wrote the passage referred to above, even though he does not mention him by name.

In modern commentaries (see for example Lenk 1982, p. 16) we find Kapp's book spoken of in glowing terms as a work ahead of its time. At the same time it is pointed out that his work had very little impact when it was published, and has only entered the arena of debate during the last few decades, when technology has become a popular area of study, even for philosophers. The book was republished in 1978.

The disappearance of Kapp's book from the academic debate is also linked to the tradition in which he was writing. Kapp must be given the credit for having put technology as a theme firmly on the philosophical agenda. (Hegel's hints in this direction were overlooked by everyone, with the possible exception of Marx, who was hardly regarded as a professional philosopher.) He did this, however, completely within the bounds of the main trends of German philosophy at the end of the last century. On the one hand the inheritance from German idealism, from Kant and from Hegel, was still making its presence felt, not least in the form of theories about human subjectivity and the history of its development. In Kapp this is most noticeable in his representation of human consciousness as self-consciousness, and his attempt to situate this self-consciousness within a general theory of development and evolution. He refers to both the individual consciousness in isolation and the idea of postulating a history of development for the human race. This need to postulate histories of development, to observe phenomena in an evolutionary perspective, was strongly influenced by the speculative, naturalistic and biological trends which had come about, perhaps in the wake of Darwin in particular, but which also had roots in the German romantic philosophy of nature.

As a generalization, intellectual history has not been kind to German philosophy in the period between Hegel and Frege (the latter did in fact challenge these naturalistic trends). Philosophers' cravings for biological speculations can perhaps best be understood as an attempt to strengthen the position of philosophy by adopting elements from an empirical natural science which was growing in importance and gaining increasing academic influence. As I mentioned in the Introduction, this was also one of the main motives behind the particular movement that was called *Technologiephilosophie*. It is therefore necessary to remove Kapp's theories from the framework within which he formulated them, in order to discover what can still be of value in them.

Kapp himself writes that the real theme of his book is:

> The origin of mechanisms modelled on the pattern of the organism and the understanding of the organism based on mechanical contraptions. In short all the carrying out of the aims of human activity through the principle known as organ projection and which only becomes possible in this way... (Kapp 1877, p.VI).

The essential expression here, and indeed in the whole of Kapp's book, is organ projection. It expresses his entire theory of technology in a nutshell. This term however contains many components of meaning, many of which we will also find in Gehlen's work, and which we must disentangle from each other. It is indeed an open question as to whether Kapp's theory is consistent.

To start with we can state that the concept of organ projection describes the basic structure of what we today would call the use of tools, the fact that human beings strengthen and extend the functions of their organs, first of all the hand, by means of objects found in nature.

> Man transcends the opposition (to an external object) because, starting from his original faculties he is able by means of mechanical support, by the work of his hands to expand indefinitely the receptive and active sensory apparatus he shares with animals (p. 25).

And further:

> Man utilises objects which are to hand in his immediate environment. Thus the first tools appear as means to expand, strengthen and reinforce the bodily organs (p. 42).

This is projection in the sense of extending and strengthening the range of the actual existant organs. A still more important element is the idea that new technical aids and inventions are created by means of *analogies* with human organs. The lens, for example, is not only a means to strengthen the function of the eye, it is itself constructed on the basis of an analogy with the structure of the human eye (p. 77 ff.). None of these thoughts are in themselves new. The historian of technology, van der Pot, traces them back to the sophist Protagoras, the Church Father Origen and the schoolman Hugo of St. Victor, (Pot 1985, §135).

Kapp thus uses the concept of organ projection in order to reconstruct the development of technology as a complete and continuous process. He does not only apply it to individual tools; also complicated machines are understood as composed of elements that are based on organ projection. At the next level these machines are then themselves perceived as separate parts of the organism, projections on ever higher levels. It is constructions like these which can be said to build on or establish vague analogies between Hegel's method in describing ever higher levels of development of the mind, and conceptions of natural selection as progress towards higher or more perfect forms. An essential stage in this development, not surprisingly considering Kapp's vantage point in history, is the steam engine, the engine par excellence and a precondition for the new historical and social conditions that arose with industrialization (p. 126). This system was stretched in an attempt to accommodate complexes like railways and the telegraph, what are today known as technological systems or networks, as well as individual inventions.

Kapp finally tries to use the principle of organ projection as the foundation for a total evolutionary philosophy of culture and theory of socialization. The entire development of mankind is conceived as an expression of ever more extensive forms of organ projection:

> ...all cultural phenomena, the crude material ones as well as those of the most exquisite composition, are nothing more than organ projection (p. 27).

In the end the simple tool has become animate and ends up as the state, as a projection of the body seen as a whole, as a concordant unit. Analogies between Hegel's philosophy of spirit and the theory of evolution become concrete fabulations about analogies between organs and institutions — we know about them in Kapp's era from the writings of Herbert Spencer. They have been assimilated into an organistic German philosophy of the state in such a way that the interesting point in the concept of organ projection as a theory of technology comes close to being lost altogether.

There is another essential element in Kapp's idea of projection, which is also almost totally concealed by the speculative framework: Kapp was aware of the fact that, by making this concept the key to his philosophy of culture, he was breaking with the widespread premise that man's surroundings were composed of an original nature. By projecting organs, in the form of models for tools, on to the surroundings, organ projection becomes a heuristic aid to sensory perception and observation, both in the direction "out from" and "in towards" the body. By means of the processes connected to the use of tools we gain increased understanding of and insight into both natural objects and the physiology of our own bodies.

> On the one hand every organ is a tool in the widest sense of the term, i. e. as a means to extend the immediate superficial perception of the object. On the other hand, the tool, the result of the work of the hand and the brain, stands in such an essential internal relation to man himself that man considers the product of his own hand as something linked to his own being...in short a part of himself (p. 25 f.).

In Kapp's writings also these thoughts are formulated within a framework which relies on both Hegel's theory of history as a process in which the mind regains itself by means of new and ever more extensive forms of understanding, and the idea of a biologically determined development in man from the unconscious to full self-knowledge. He is operating with the idea that organ projection is based on an insight into the structure of the body that is subconscious at first, but of which man becomes aware by means of this objectivation.

What is original in Kapp's work, an element that breaks the mould of the philosophy of the spirit (*Geist*) or the speculative framework, is that he associates this process with the human body and its organization. Organ projection thus becomes, as we have already seen (see the quote above from p. 25 f.), a process which contributes to man's understanding of himself, in that the original projections of the organs, the tools, are the key to an understanding of their original patterns, the organs. Kapp calls this the anthropocentric point of view and refers to Protagoras's *homo mensura* principle (p. 13).

Today it is more fitting to say that Kapp's anthropology anticipated themes in modern philosophy, also within what I have called materialistic phenomenology. An important theme in phenomenology is precisely this question of how the human subject is situated in the world. In Husserl's theory of the lifeworld, Heidegger's teachings about *Dasein*, and in the French version of phenomenology we find various attempts to describe this relation, attempts which all share the common aim of trying to escape from the subject–object distinction in traditional epistemology. By way of contrast to, for example, Heidegger, Merleau-Ponty emphasized in particular the fact that we are in the world as *bodies*. The experience of being in a three-dimensional space, for example, is due to the fact that we are projecting the construction of our own body on to the world.

But what meaning could the word 'against' have for a subject not placed by his body face to face with the world? It implies the distinction of a top and a bottom, or an 'oriented space'. When I say that an object is *on a* table, I always mentally put myself either in the table or in the object, and I apply to them a category which theoretically fits the relationship of my body to external objects. Stripped of this anthropological association, the word *on* is indistinguishable from the word 'under' or the word 'beside' (Merleau-Ponty 1962, p. 101).

It is this kind of philosophy about the body and about a body-subject which Merleau-Ponty later developed in full and which Kapp anticipates in his theory of organ projection. As is also shown in his repetition of themes from Marxism, Kapp was in spite of everything first and foremost a post-Hegelian.

In my discussion of Kapp I am confining myself to his theory of organ projection as a philosophy of technology, as a theory about the origins of tools and machines, and their logic of development. I will postpone dealing with the question of the relevance of anthropological theories of technology as theories which can explain the development of civilization and socialization until I have presented Gehlen's theories.

As far as I can see, two main objections can be raised to Kapp. He claims in response to Humboldt that scientific apparatus must be regarded as *new* organs (Kapp 1877, p. 105). One is tempted to ask whether this view, which is reasonable enough when considered in isolation, does not weaken the theory of projection rather than support it. We seem to need, as Gehlen shows, a theory which does not only deal with projection, i.e. extension and reinforcement, but also with the *relief* of organs. Of course it is possible to claim that organs are relieved by means of tools and instruments modelled on analogies with *other* organs. Nevertheless it seems rather bombastic to claim that all technological elements are modelled on some kind of organic pattern; why shouldn't observations of other natural processes and natural phenomena, such as the behaviour of animals, also have provided inspiration for technology?

More serious is a problem I mentioned in the Introduction: One cannot consider all technology as mimetic. For example, such an important factor in the development of technology as fire and the uses it can be put to, which seem only to be found in the species Homo, has no organic model in humans or animals. Of course the use of fire may be linked to observations of its effects, which are then reproduced in controlled forms. But this is hardly mimesis, and most certainly not organ projection. Even the technology of tools, taken as a whole at any rate, is not really mimetic, as it is not linked to, or exclusively linked to, the reproduction of patterns given in nature. The most interesting example of this is the use of technology which depends on the control of rotating forms, inventions such as the wheel and the turntable. These objects have been developed without models in either the human organism or organic nature. Of all the things that natural selection did not produce, the wheel is the most striking example (Gould 1984, pp. 158 ff.). Technology based on rotation is difficult for humans to control and appeared relatively late in the development of human society.

This is even more true for modern technology. The aeroplane may be the realization of an ancient dream of flying *like the birds*. The principles of heavier-than-air flight are not however analogous with those that control the progress of birds. It was only when we gave up imitating both the beating of wings and gliding, and moved on to experimenting with fixed, rigid wings and gained access to a source of energy that had a large enough output in relationship to its weight, that the *aeroplane*, i.e. a machine that had mastered heavier-than-air flight, became possible.

# 2 Arnold Gehlen

Gehlen's theory of technology was developed as a part of a project which aimed to provide a total theory of man. This theory has two main parts. The theory of technology is thus supplemented with, or rather becomes a part of, the theory of the institutions of human society and their necessary and inestimable function in maintaining society. Right from the start, however, these two parts are inextricably connected to each other through certain interpretations of the basic characteristics of human nature. Gehlen uses the doctrine of institutions as his starting point for an acerbic criticism of contemporary culture and the role of intellectuals: In our epoch the permanent institutions have been weakened and we seek — of course to no avail and with catastrophic results — to replace them with an unfettered reflection and adoration of the purely subjective elements in our consciousness. Among the liberal admirers of Gehlen, especially in the USA, there is a tendency to try to see Gehlen's political and critical polemics of contemporary life as independent from his more neutral anthropological theory. By contrast his German critics, Jürgen Habermas in particular, lean towards the opinion that Gehlen's theory of socialization and therefore also his theory of institutions have inherent, decisive weaknesses and flaws which in fact pave the way for his conservative and anti-individualistic condemnation of modern, subjectivist culture.[3] Personally I lean towards this last interpretation, as will become clear at the end of this chapter. But this does not imply that it is fruitful to start off by passing political judgement on Gehlen, and certainly not on his theory of technology. In fact Habermas, in his early work, goes further in his support of Gehlen's theory than I do (Habermas 1971, p. 87).

The central concept in Gehlen's anthropology or complete theory of man is *action*. The reasons for this are both systematic and historical. In the systematic sense it is a matter of surmounting all the previous attempts to solve the mind–body problem, attempts which, according to Gehlen, have all ended in a deadlock.

---

[3] See, for example, Peter Berger's "Foreword" to Gehlen 1980. For Habermas's criticism see the end of this chapter. In Lenk 1982, p. 270 ff., there is a comprehensive critique of Gehlen, but it deals in the main with different aspects of his writings to those I am examining here. Lenk claims that Gehlen lapses, against his will, into a form of idealism, that he dissolves nature external to man in systems of interpretation that are culturally dependent.

The starting point we seek is *action*. The deliberately performed human action is as a process in its real course a preproblematic entity which cannot be dissolved into different experiential elements (*erläbnissmessig völlig untrennbare, vorproblematische Einheit*). The task may now be described thus: To construct a general anthropology on the basis of action. We must lay down some definitions: By *action* we are to understand the forward-looking, planning alteration of reality and the sum total of the facts which have been altered or produced in this way, as well as the necessary means for these operations, "ideal" as well as "objective" means, are to be called *culture* (Gehlen 1942, p. 71).

The systematic reason for approaching the problem from this angle is then basically to avoid all dualisms between the external and the internal, between mind and body.

This point of departure can of course seem almost arbitrary. Gehlen's choice of precisely this point of departure may become more obvious if we look at his retrospective view. He presents his orientation towards philosophical anthropology first and foremost as a break with ontology as the basic philosophical discipline. In interwar Germany it was possible to a certain extent to lean upon Max Scheler's anthropology, especially his emphasis of man's "*Sonderstellung*" (special situation) in relation to nature. But to understand action as a basis for this special situation was, according to Gehlen, a break with Scheler and a movement towards American pragmatism (and perhaps a result of the influence of Marx that he could not or would not admit to (Gehlen 1968, p. 210).

In other works Gehlen reaches further back than the interwar period to find the background for his thinking. He gives prominence to Fichte as one of the most important theoreticians of the last few centuries, underlining in particular his description of man as a creature that has liberated itself by regaining the independence it has invested in its products, a development which can be described as objectivation, alienation through objectivation and sublation of this alienation. Here we encounter an essential element of Gehlen's criticism of culture, for he adopts from this dialectical tradition the very idea that the existence of man necessarily implies objectivation and to a certain extent alienation. At the same time he rejects the idea, which was equally central to this tradition, that it is possible or positive to regain a lost freedom, even if it is in a new form, by doing away with those features in the formation of society which lead to alienation. The internal dynamics and objectivity of the institutions must remain as overindividual elements; they cannot be mediated with subjective pretensions of freedom, as Hegel had hoped. Only this kind objectivated situation can provide us with the security we need. The need to remove alienation is a neurotic attempt to replace control and order by freedom (Gehlen 1952b).

Let us now however pick up the thread that we dropped earlier by querying Gehlen's sources of inspiration. The decision to make action a basic concept has its origins in Gehlen's view of man's place in nature. A creature with the kind of biological properties that man possesses can only survive by means of a reorganization of nature for its own purposes. This is the idea behind his claim that we have to see *action* as central to all further problems and questions (Gehlen 1941).

These basic features thus characterize man as one biological species among many. But they also make man so different from all the other species that it is pointless for Gehlen to view man as merely a biological species. Man cannot really, or if you prefer, can only negatively, be described in biological terms. These basic characteristics can be summarized as follows: Humans lack specific organs that are capable of adapting them to a life in a set biological niche. They lack specific fighting organs and protective features. They are not equipped to react instinctively to features in their habitat, e.g. the arrival of an animal that is their typical prey, or for which they are prey. Gehlen maintains further that man is a "late developer", he grows old more slowly than other animals, and the basic characteristics of an adult human retain some childlike features. All this points to the retention of features from the foetal stage.[4]

To summarize, man is not very evolutionary specialized. On the contrary man is in fact *unspecialized*. Gehlen introduces the concept of man as a "incomplete creature" (*Mangelwesen*), which in turn corresponds to the state of being "open to the world" (*Weltoffen*). These terms have been borrowed from the German Romantic Herder and Max Scheler respectively, the latter of whom can be regarded as the founder of the genre philosophical anthropology.

Gehlen's choice of action as the basic concept is at the same time also a result of the fact that man, almost by definition, must be understood as a being that creates the environment in which and on which he lives. Gehlen reserves the term surroundings for what we today would call the ecological habitat of a species. Man has no such specified habitat (Gehlen 1942, p. 79). Culture according to Gehlen is the exact opposite of adaptation to the given

---

[4] This view, which originates from the Austrian biologist Bolk, has not been entirely rejected. The phenomenon with which it deals, that man retains some juvenile characteristics and matures more slowly than other mammals, is called neoteny. The theory of neoteny is defended by, among others, the prominent modern biologist Stephen Jay Gould. He combines it with the standard theory of evolution by showing that it also holds true, to a certain extent, for primates (Gould 1978, p. 63ff).

surroundings. This is because of the fact, as we saw earlier, that the essence of man is that he is biologically underspecialized, he lacks the basic, instinctive mechanisms of adaptation to ecological habitats (Gehlen 1941, p. 55 and Gehlen 1980, p. 3).

It should already be clear how fundamental a position the theory of technology occupies in Gehlen's complete conception of man's situation. To shed even more light on this we can take a look at Gehlen's view of biological theories about man.

Gehlen rejects Darwinism, presumably on the grounds that he believes that an evolutionary theory of man would be irreconcilable with the theory of man's open nature (Gehlen 1988, p. 110). If Gehlen had had to stand or fall on this, then he would undoubtedly have fallen. But it is perfectly possible to conceive of a biological evolution which would leave man at a level where principles other than variation and selection steer his development. When talking about the biological determinacy of technology we must distinguish between two senses of the word determined: Firstly the specific genetic determination of any given technology, and secondly the general biological determination of types of behaviour. In order to understand the biological foundation for the use of tools in early humans, we need only to refer to the last type of mechanisms (Beck 1980, p. 140). It is advantageous to reformulate Gehlen's theory so that his description of man starts from the level where a conception that is linked to specific determination becomes irrelevant. From this point on there are many possibilities for a development of human society that no longer can be grasped in categories taken from the theory of natural selection. Without wanting to sound too paradoxical, it is even possible to say that Gehlen's interpretation of human nature is such that it makes self-sufficient biological explanations of human behaviour impossible. His extensive use of biological theories leads inevitably to this conclusion. Due to his biological incompleteness man is doomed to culture, or if you prefer, doomed to being a cultural creature.

Despite all his praise of and use of biological theories, Gehlen therefore breaks at the decisive point with what we often regard as biological explanations of man, and also with the aggression theoreticians like Konrad Lorenz. In man (in this context not only the species Homo sapiens, but also early humans) there is no original natural state (*Wildform*) determined by biologically adapted instincts. I do not know if Gehlen ever commented on the latest trend in biological explanations of human behaviour, namely sociobiology, but it is reasonable to believe that he would have rejected it. Up until ten years ago it was still usual to link biological versus cultural explanations in sociology with a political right–left scale. The fact that it nevertheless seems reasonable to interpret Gehlen as a political right-winger, as indeed he too perceived himself, must be due to features of his explicit theory of culture, and not any biologism. It was after all with reference to his rejection of Lorenz that Gehlen formulated the slogan "Back to culture", and made one of his most violent attacks on modern subjectivist culture (Gehlen 1952a, pp. 132).

However, whether Gehlen implements this approach consistently is a different matter. Every now and again we notice in his writings that certain features of technology or man's relationship to his institutions are traced back to fundamental characteristics or instincts. On the one hand this allows Gehlen to introduce elements of Freud's instinct theory into his own theory. On the other, the status of these essential characteristics, which he introduces quite at his convenience, becomes extremely unclear.

We can now turn our attention to Gehlen's theory of technology proper. Like Kapp's theory, it concentrates on the relationship between organs and tools, but Gehlen's view of this relationship differs somewhat from Kapp's. He operates with three forms of technical skills: *Reinforcement techniques* are added to and extend the skills and abilities we already have; *Compensation techniques* enable us to perform operations for which there is no "basis" in our organic make-up; and finally *Relief techniques (Entlastungstechniken)* reduce the amount of energy needed and liberate the organs. The result of and the means of these forms of technical skills constitute human culture, or our "second nature".

Gehlen overcomes some of the weaknesses in Kapp's theory: It is of course possible to say that we relieve our organs by means of tools which are based on projections of them, but Gehlen's concept of relief is nevertheless broader than Kapp's concept of projection. Gehlen is well aware that technology is not mimetic and uses the wheel as an example of an element of technology which has no organic model. In his view, technology strengthens our abilities or gives us skills we do not have, but it is not modelled on things within or without us. In this area too, however, one can question just how consistent his interpretation is. According to Gehlen (Gehlen 1980, p. 4), who well knows that the aeroplane is not based on mimetic technology, all three principles can be found in the construction of the aeroplane. We gain wings that we do not have, and which cannot be compared to those of any bird, and we are able to move without effort. But there is no sign of any strengthening or relief of *our* organs here. By way of a summary we can say that Kapp's starting point is insufficient because not all technology is based on projection, and Gehlen's because it is difficult to determine which organs are relieved through modern technology. We are now seeing radically new technological developments which perform new functions that our organs cannot even begin to fulfil and from which they therefore cannot be relieved either.

Gehlen's general strategy should nevertheless now be clear. The principle of relief functions without a doubt as a kind of metaprinciple for technology. As an extension of the process where our organs are extended and strengthened we are also relieved of direct motor activity. This objectification of the external world leads inevitably also to an ever increasing relief. This is the only way in which man's need to secure and stabilise his surroundings can be satisfied (Gehlen 1980, p. 17). The meta-character of the principle of relief is expressed through the very fact that it is this saving of energy in labour, that leads in the first instance to increased control over the world, and later also to an objectivation of man himself, via the appearance of routines and the formation of habits (p. 18). The significance of the principle of relief is as such a direct result of the fact that man is an incomplete creature (*Mangelwesen*).

Through the relief of organs rendered possible by technology appears a field, call it culture or second nature, where we do not relate to direct external influences via our senses, but are capable of planning and predicting. We relate to expected and future results and events, not to the actual occurrences. In a certain sense, we do not need to relate to the latter as they occur with a reasonable degree of certainty and expected regularity. This is, as we have seen, an essential component of meaning in the idea of relief. The application of technology does not only lead to social institutions; technology has *itself* already assumed the characteristics of institutions. It has created predictable ways of reacting and responding in us, or, to put it another way, Gehlen attempts to understand institutions on the basis of the same features — man's lack of biological specialization and his character as a creature of action — that he uses to deduce the significance of technology. His task is to deduce:

> ...from human nature the uncoupling and autonomy which the institutions obtain in relation to the individual (Gehlen 1977, p. 8).

Although Gehlen does not subscribe to Kapp's theory of organ projection, we nevertheless find the idea in his work that technology may be used to help us understand our own organism. The steady rhythm of both mechanical processes and natural processes gives us a clue towards understanding our own physiology. On the basis of such thoughts Gehlen believes he can introduce the concept of technology as a phenomenon of resonance, a perceived agreement between technology and instincts, (Gehlen 1980, p. 14).

Even on the extremely general level on which Gehlen is operating, these observations have significance for modern societies with rapid technological change. I claimed above that Kapp's and Gehlen's theories of technology are best suited to describe a very primitive stage in the development of human society, and cannot really be used to explain the process of technological development as a whole, if there is such a thing. I must however concede that the idea that interaction with technology is characterized by relief through formation of habits and routinization, a virtualization of consciousness and knowledge in such a way that technology both relieves the amount of knowledge and ability we need and at the same time leads to new processes of formation of needs and preferences, is probably more appropriate for technology in the twentieth century than in the eighteenth century. Here are two significant points:

First of all, Gehlen's theories shed some light on the phenomenon known as technological alienation. In the literature of criticism of technology there are numerous examples where certain technologies are described as decisive steps along the road to cultural decay and increased meaninglessness. At the same time these same authors take for granted those forms of technology with which they have grown up and do not see them as problematic, whereas the technological advances that are new to them and to which they are now reacting negatively have long since been accepted and incorporated into the field of unproblematic habits by the next generation. Both phenomena are understandable — nothing creates greater alienation than finding that one's own old routines have become dysfunctional or meaningless. This proves, however, that this kind of reaction is not founded on the fact that technology has been introduced into a new field for the first time, but rather that a new form of technology has replaced an old one. Technology leads to what we can call secondary traditionalization: Its application has become unproblematic. This does not mean that technological alienation in this sense can be seen as a merely generational phenomenon. This is of course part of the picture if we give up on the idea of the existence of a natural technology, or at least that this presumed natural technology is not identical to the technology with which we ourselves have grown up. Such conflicts are concrete expressions of the complexity inherent in the problems of modernization, the relationship between technological imperatives and cultural transmission, the relationship between the various groups of actors such as users and designers, those affected and the decision-makers. This is not made obvious, however, if we insist on clinging on to the abstract opposition between technology and culture, instead of talking about contexts of the lifeworld adapted to various forms of technology.

Secondly, the gap between the layman's actual use of technology and artefacts, and the technological and scientific competence and knowledge invested in them, is wider today than ever before. Steam engines were primarily not only constructed by, but also operated by experts, even if not by scientists. Electricity, however, has opened up the field for apparatuses that everyone can use and which are linked to application and communication contexts of the lifeworld, but at the expense of our being relieved of closer insight. In the age of the word processor and the Pill this pattern has developed even further.

This is probably part of the reason why today there is a more widespread, diffuse uneasiness about technology in modern societies than previously, that cannot be subsumed under the concept of technological alienation as I have been using it here. It can probably *also* be traced back to the fact that we lay-people come into direct contact with and ourselves use forms of technology that are unknown to us to a much greater extent than before, rather than only to the fact that there are now uncontrollable forms of technology. Of course we can find counter tendencies here: There are traces of gradual hierarchies between laymen and the learned, the best example of which is the "hacker" of the computer world. Computer technology is itself an example that disproves the theories that claim that all education will be formalized in the technological world. In this field we are facing a new wave of learning by doing and informal training at the workplace.

Despite the fact that Gehlen's theories of technology help to illuminate these connections between theoretical insight and application in the lifeworld, we are now moving into an area of which Gehlen himself was unaware, and which probably also marks the limit of the capacity of anthropological theories.

I shall conclude this chapter by 1) considering if such theories provide an acceptable interpretation of the role of technology even in societies with primitive technology; 2) considering whether technological change in the perspective of the history of civilization can be explained by such theories; and 3) considering whether these kinds of theories can perform their intended function, i.e. as general theories of social development and of the processes of socialization.

1) First of all, Gehlen's analysis of the use of tools is insufficient, even at the simplest level, such as that which we find in animals and early humans. According to Beck (Beck 1980, p. 190) the use of tools has four main functions: 1) to increase reach; 2) to increase mechanical strength; 3) use in social contexts such as games, and to scare others; and 4) control of liquids. Gehlen's interpretation covers only functions 1 and 2 in this more detailed picture. It is also doubtful whether it is correct to view all aspects of societies with simple technology as inextricably linked to the society's technology as Gehlen claims. This view would appear to have been handed down to him by an out-dated anthropology, and it is interesting to note that Gehlen in this

way is subject to the same criticism which can be levelled at one branch of Marxism, namely the tradition stemming from Morgan, especially as it manifests itself in Engels (Engels 1968).

In this tradition all social relations, such as relationships of property and kinship, were perceived as closely correlated with the technology in the specific society. In the following I follow Maurice Bloch's criticism of Engels (Bloch 1983) which I find convincing. Also Gehlen's attempt to connect technology to development of institutions implies, like Engels's interpretation, an:

> ...unavoidable link-up between technological systems, economic systems, kinship systems etc. In fact, recent findings make this kind of view untenable... (Bloch 1983, p. 64).

Bloch dismisses the existence of any definite connection between technology and systems of kinship, and therefore also between technology and types of marriage, and between technology and the status of women in various societies with primitive forms of technology. This view is supported by Hallpike who shows that it is not possible to find anything other than banal statistical correlations between technology and social structure (Hallpike 1986, chap 4).

Now Gehlen would hardly make this kind of claim. Bloch's interpretation does however imply that it is not possible to explain the various forms of division of labour, which in societies based on kinship relationships to a great extent are a division of labour between the sexes, on the basis of the technology in that society, and this turns the relationship between institutions and technology around in relation to Gehlen's theory.

> Among many groups of hunters and gatherers, we commonly find an elaborate and unequal division of labour, especially between men and women, which the technology in no way causes. In fact, in these cases the division of labour seems to be the product of the exploitation of women, rather than the cause (p. 86).

2) Gehlen states explicitly in several places that it is possible to understand the development of technology on the basis of certain fundamental instincts which are decisive factors in the whole development of technology. These instincts firstly comprise of coordination of and interaction between the senses and the organs on the basis of the principle of feedback, which he calls the circuit of action, and secondly the principle of relief, which we have already come across. The form of periodization he gives, which is based on a work by Hermann Schmidt, remains within this framework (see Schmidt 1953 quoted in Gehlen 1980, p. 16). In the *tool* period the physical and mental effort is performed by the subject itself with external reinforcements

of the organs; in the period of the *work machine* power is objectified outside the subject in such a way that the organic materials and energy are replaced by inorganic ones; and in the *automaton* period also the mental input, the actual circuit of action, is objectified. From a level of organ substitution we have attained a level of organism substitution, a movement from the external to the internal, a pattern which is reminiscent of Kapp's outline of development from the projection of individual organs to the projection of the organism as a whole. This process has followed, and by and large determined, the history of mankind, even if until modern times it was expressed through magical practices and not through technology. As regards history as a whole, Gehlen is then only in a diminished sense a technology determinist, but he seems to be one without reservation for the modern period.

All the same, Gehlen's periodization is still dubious: It cannot account for what we today perceive to be the first decisive break in the history of man: The transition at the start of the early Neolithic period from a society of hunters and gatherers to an agricultural society that grew crops and kept domesticated animals. This decisive leap in the development of human society would be included in his first period. Like so many other typologies in the history of technology it suffers from a fundamental lack: The different periods are not defined on the basis of a single, consistent theory. The defining characteristics of the various periods are on completely different levels.

Gehlen himself does of course point out that this kind of theory cannot explain the development of, for example, individual machines. In order to do that we have to draw on specific scientific and technological theories. He is also aware of the fact that science, technology and industry together are integrated into a common complex today.

It is difficult to see, however, how he accepts the consequences of this. He claims to define both the development of technology as a uniform phenomenon and the fundamental steps in its development on the basis of the same general level of postulated instincts. I say postulated for several reasons: Gehlen neither provides any reason for how they arise nor does he attempt to explain them in any other way than by claiming that they can explain a certain development. Nor does there seem to be any particular connection between his and Schmidt's periodization and these instincts. The postulations of the instincts seem ad hoc, and the theory is not compatible with Gehlen's insistence on man as a cultural being, rather than an instinctual being. These instincts seem to be forms of action on such an abstract level of social theory, that it is difficult to see how we can find evidence here of results of an evolutionary process. An attempt at a consistent interpretation of Gehlen, if one is to combine the many invocations of instincts as a basis for technology with the fundamental rejection of the concept of instincts (see Gehlen 1988, p. 321), would be to maintain that instincts are to be understood as biological impulses which have already been captured by the process of socialization and thus have been isolated from immediate satisfaction. They have become connected to other goals than the original ones. But even this kind of interpretation cannot furnish these impulses with a position as the trigger for and supporter of the process of development of technology as a whole, and even less serve as a basis for a periodization.

3) These kinds of misconceptions are not only manifestations of the inevitable limitations which affect all general theories of society that build on empirical data. They indicate a fundamental weakness in regarding theories about technology as a basis for theories about processes of socialization, and this is precisely what the anthropological theories are, in Kapp, Engels and Gehlen.

Kapp's application of the principle of organ projection is of course pure speculation. But also Gehlen's more serious linking of technological and institutional development, his identification of culture with technological processing of nature, exhibits an inherent bias. Seen in this light, his antimodernism is not an accidental and purely personal extension of the original approach.

I will limit myself to Habermas's criticism which can be summarized in two points.[5] Gehlen launched the slogan "A person is an institution in a concrete case". This is of course the natural consequence of drawing out the idea of relief and formation of routines to its logical end. This means that Gehlen must miss out on the moment of individualization, of development of personality linked to *both* the ethical autonomy in universalistic ethics *and* the individuality-connected, expressive and aesthetic *"Selbstdarstellung"*, which both seem to be necessary moments in our modern self-understanding. Gehlen interprets individualization as the origin of the individual subject only as far as numerical identity is concerned, not as the origin of individuals who are qualitatively different to one another (Habermas 1988, p. 187).

The second point is this: According to Habermas in a different context, Gehlen's theory of socialization is unable to account for the intersubjectivity which we have presumed came before the formation of institutions, and which is interlocked with language as the mechanism of both individualization and socialization. On this level, patterns taken from the processing of physical objects are not in fact *patterns*, no matter how much — and I add

---

[5] For an early version of Habermas's criticism of Gehlen, see Habermas 1958. The entire relationship between Habermas and Gehlen is discussed by Wilhelm Glaser (Glaser 1972), but his book lacks interesting views on the subject of technology in the narrow sense, although it does contain some more innovative ideas about the technocracy thesis.

this at my own risk — they are able to contribute to grasping these processes of socialization (see Habermas 1981, p. 118).

Gehlen's break with nature thus appears to be more incomplete than mankind's own.

# Part II
## THE DIALECTICAL TRADITION

# Introduction to Part II

When looking at the anthropological philosophy of technology we encountered a way of thinking that first and foremost moved within the framework of an interpretation of man as a tool-user, a creature which itself creates its own surroundings. As I mentioned earlier, anthropology is defined here as philosophical anthropology in the sense attributed to this concept in interwar German philosophy. We were confronted with interpretations on a high level of abstraction, far removed from empirical field work. The most useful concept that this tradition has given us, a concept that can supplement the concept of materiality, is the idea of technology as *relief*. To a great extent, technology is the process by means of which functions of the human body are transferred to operations of and with artefacts. In Gehlen's philosophy we were presented with his idea of the relief of physiological operations according to a system which was then dubiously extended to cover the whole history of human society. In connection with Habermas and Luhmann in Part III we will return to this theme with regard to the question about the possibility for and limits of such a relief of symbolically mediated communication.

Within the dialectical tradition I would like to start off by returning to questions concerning epistemology, as perceived by the German philosophical tradition in the period after Kant.

This kind of perspective may appear to be a step back to even more abstract interpretations than those we looked at in Part I, and thus ill-suited as a starting point for a concrete discussion of technology in modern society. But it turns out that the tradition from Kant, via Hegel and Marx, up until Sartre, in fact provides an opening for the very discussion of technology in a historical perspective that is missing from the anthropological approach. This special approach appears in a philosophy which deals with *labour* rather than technology directly, and labour as an important, albeit not the fundamental factor in man's relationship to his surroundings. This is the reason why I have chosen not to use a strict chronological account in this book as a whole. In Part III I am then going to confront also the dialectical tradition with Habermas's criticism, in order to see which of its results can be developed further on a new basis.

# 3 Georg Wilhelm Friedrich Hegel

By way of an introduction we can emphasize two elements of the situation with which Hegel was confronted through Kant's epistemology, and the continuation of this philosophy in Fichte and Schelling.[6] Kant broke with the traditional and absolute distinction between a passive sensation (or receptivity) and an active ability to think or reason which had dominated European philosophy both in its various metaphysical forms and in the subjectivity-oriented philosophy which arose in the early modern period. The theory of active reason as the real human contribution to knowledge stems from late mediaeval nominalism (Blumenberg 1973, p. 227). The theory that the distinction between sensation and thought, between sense-impressions and conceptions, was rooted in the passive nature of sensation survived all the feuds between the rationalists and the empiricists in modern times. A view which seems to have dominated the metaphysical tradition may be formulated thus: To sense is to apprehend concretely and materially; to think is to apprehend intellectually and generally.

Kant broke with this view in several ways. Via his doctrine of the a priori forms of intuition in our sensibility, he paved the way for a new concept: Sensible intuition. The concept of sensible intuition includes elements from the traditional interpretation of both senzation and conceptual knowledge without fitting into any of the traditional categories. An intuition in Kant's sense of the term has a structure which goes beyond what we find in isolated sense impressions, but the apprehension of something by means of sensible intuition is not to apprehend this something by subsuming it under conceptual characteristics.[7]

Let us look at an example: According to the traditional interpretation we receive isolated impressions through our senses, such as yellow, heavy, hard. When we conceive of an object through concepts, we define the object by the qualities hardness, yellowness and weight. (There was of course much discussion as to how far one would get in one's understanding of nature by merely building on these kinds of qualities which were provided through the senses, as is made clear in Descartes' famous wax example in the *Medita-*

---

[6] This chapter is a heavily revised version of my PhD thesis (Krogh, 1985).
[7] It would go beyond the framework of this thesis to justify this interpretation of Kant in detail. The relevant parts are explained in my doctoral thesis, (Krogh 1985), part I in particular. For an explanation of the concept of intuition below, see Sellars, W. 1967.

*tions*). A sensible intuition of a piece of gold, by contrast, comprises neither isolated impressions nor the perception of the object by subsuming it under the concept of gold, but that what we perceive becomes part of a structure that goes beyond what can be contained by the traditional theories of sensation. It fills a spot in a three-dimensional space in relation to other objects and our experiencing of it fills a period of time in our context of experience, where this experience comes before or after other experiences.

Kant also appears to claim that these structures in our consciousness are ultimately due to an activity in our thinking. The distinction between sensing and thinking, understood as the distinction between passivity and activity, is broken down by moving the spontaneous activity of understanding into sensing. It is this point that makes it possible to connect theories about labour, and therefore also theories about technology, to this purely epistemological problem: Thus a theoretical place appears for an incorporation of labour into an epistemological context. The concept of sensible intuition also paves the way for another concept: Labour can be conceived as *active* or *productive sensation*, or to put it in a more Kantian way, active *intuition*. Through this theoretical definition of labour we can then introduce the concept of social materiality into the dialectical tradition.

It is of course possible to interpret the development of epistemology up until Hegel according to a somewhat different model, which was developed by Ernst Cassirer and later by Georg Lukacs. It even struck a chord with Marx. This interpretation has its starting point in Vico and his formulation "Verum, factum est". The only reality, or in Lukacs's words, the only form of objectivity we know, is the one we ourselves have produced. For Vico this was first and foremost the historical context of tradition, but this view appears to become more and more fundamental to the apprehension of all objectivity in the tradition we are dealing with here. This principle can thus be understood as an expression of an increasing consciousness about man's world-producing function. If we ignore the speculative nature of postulating this kind of connection between epistemological development and actual history, which we find in Lukacs, then my somewhat more modest perspective, which starts with Kant, appears to have the advantage. It directly accentuates the connection in Kant between that level of activity through concepts of reason which he calls transcendental synthesis, and the level of sensation, and it is precisely this connection, which is implied in the concept of formal intuition, that is expressed in the idea of an active *intuition*.[8]

I am hoping to show how the set of problems I have hinted at here are expressed in Hegel's writings. The following comments do not of course

---

[8] For this interpretation see Krogh 1985, especially chapters 10 and 11, and the literature referred to there. The relevant passage in Kant is Kant 1787, p. B160.

pretend to be even a brief summary of Hegel's basic presuppositions. For example, I will not be looking at the special problems associated with his logic and system as a whole. Fortunately this kind of total interpretation is not really necessary in a book about technology. I shall therefore concentrate on looking at how labour, and therefore also the technology of tools, appear in Hegel as a (single) moment in the development of the subject and its relationship to its surroundings.

Charles Taylor's interpretation of Hegel provides a very useful starting point for an understanding of both Hegel's place in the growth of a philosophy of labour, and the role labour plays in his philosophy (Taylor 1975). He describes Hegel's basic position as "expressionism". Expressionism is taken here as a theory or rather a handful of theories about the nature of man, and his relationship to his surroundings which Taylor traces back to Aristotle and Herder. In expressionism the Antique idea that man is the result of a process of realization is coupled with the modern conception in Romanticism that man is not any one given thing, not even given as one particular possibility for one specific result. Man is his own product in the radical sense that he not only realizes himself, but it is only his own activity which is involved in this process. Self-realization occurs by means of various media, such as language, labour, art etc. In the vocabulary of expressionism, man is then conceived of as a being which objectivates itself by externalizing itself, and through this process also creates a world. In this context it is important to emphasize that in externalization in media such as labour and language there is no set or definite being in man which is externalized; it is created in the same act as its surroundings and is transformed in the same way and to the same extent. Taylor thus perceives the role of labour in Hegel's writing as one of the media which man uses to simultaneously transform his surroundings and thereby produce himself, and his interpretation includes Hegel's concept of labour in its key concept.

Taylor writes:

And this brings us to a second important function of work: in transforming things we change ourselves. By creating a standing reflection of ourselves as universal beings we become such beings (Taylor 1975, p. 156).

This quote clearly expresses the interaction between subject and object, and the simultaneous and mutual forming of both elements.

I am not going to discuss here whether Hegel's entire philosophy ought to be understood on the basis of labour as its basic medium, as first Marx (Marx 1969a) and later also Lukacs (Lukacs 1973, vol. 2 chap 3) have claimed, or whether Hegel wavers between, on the one hand, distinguishing labour and language from one another, and on the other combining them into a higher entity, so that the typical intersubjectivness of language is lost, as Habermas claims (see "Labour and Interaction" in Habermas 1974). If one were to constrain oneself to a perspective of purely Hegelian philology then Habermas is probably right. In the medium of labour we remain on the level of a subject–object relationship, and it is not in keeping with Hegel's presuppositions to remain attached to this kind of perspective.

I am of the opinion, however, that the term expressionism may be misleading on one point, namely as regards underlining the two-sidedness of the forming relationship; the medium of labour produces both new objects and new types of human beings. It is also clear that Taylor's presentation is somewhat indecisive in places. When introducing the term expressionism (Taylor 1975, p. 81 ff.) he speaks of self-realization in such a way that the element of self-creation is lost. The same is also true of his discussion of the relationship between Hegel and Marx (p. 129). It may be correct that Hegel primarily saw labour as a means for changing man, while Marx, under the influence of the industrial revolution, attached more importance to the changes to man's environment. It is however problematic to turn at any rate Marx's conception of labour into a theory about a realization of man's being in the Aristotelian sense. This would be overlooking the depth of the idea of the alienation of man in Marx's theories of labour.

But regardless of this, we can still agree with Taylor's definition of the function of labour in Hegel's philosophy:

> [...] man can come to see himself in the natural environment by making it over in conformity with his own project. For in doing this we achieve another standing negation, a reflection of ourselves which endures (ibid.).

Let us now take a look at how these thoughts are expressed in a couple of Hegel's early works and outline the details of the way in which we can view the tool as the medium for a reciprocal production of subject and object.

The first of these works is the so-called *Jenaer Realphilosophie* from 1805/6 (Hegel 1932). Until an ultimate and reliable edition of the complete works of Hegel is published, it will not be possible to say with complete certainty which philosophical system Hegel had in mind in connection with this work. Although labour is treated later in a different context to the one in this manuscript, so that we are walking on thin ice here, it nevertheless tells us something about a stage in, and therefore also about the development of, Hegel's thoughts on labour and technology.

In *Jenaer Realphilosophie* labour is discussed in two parts: Firstly in the paragraph on the "subjective mind" where it plays a part in the awakening of the mind from the animalistic state, in the genesis of the conscious individual; and secondly in the paragraph on the "real spirit" in connection with the dialectics of acknowledgement. In *Jenaer Realphilosophie,* then, labour is placed within a philosophical theory about the development of man from the animalistic to the social state, from the perception of images, via naming, to labour in modern society with all its special social and technological assumptions, a development which later dissolves into categories that become ever more intersubjective. Typical of these relations is what Hegel calls love (*Liebe*), a category for immediate human interdependence which he used in his criticism of Kant's "formalist" ethics. In this last context, labour is placed in connection with the principle of division of labour, a field where Hegel is generally content to follow the theories of Adam Smith. Here Hegel anticipates the topic which later, in his *Philosophy of Right,* would become "the system of needs" (Hegel 1821, §§189–208). At the same time we also find a clear indication that Hegel had understood how the process of production that was based upon the division of labour would necessarily seem alienating. His representation anticipates in truncated form the picture that Marx painted with broader strokes in his early writings.

Labour then is depicted in the first part as a human mode of activitiy which is linked to a stronger ability for establishing and adhering to general rules than that which we find in a purely linguistic contact with the things. What we are actually dealing with in this sketch is a description of a progression in the ability to grasp, maintain and apply general rules. Hegel is basing his ideas here on a theory of language which makes naming, and not the ability to form meaningful utterances, the fundamental aspect of language, a theory which in our post-Wittgensteinian times has generally been discarded. This means that his treatment of the relationship between labour and language must today be deemed somewhat antiquated, although this is not necessarily also true for his views on labour seen in isolation.

Basically we can say that labour is the ego's or the subject's conquest of itself — through its conquest and transformation of the external objectivity, through labour the subject first of all makes itself an object:

> This labour is thus the first inner influence on itself, a completely non-sensible activity and the start of the ascension of the spirit (Erhebung) — because here it is its own object (Hegel 1932, p. 188).

Labour here causes the subject to attain a relationship to itself through a relationship to another being and a liberation from thingness through just such a forming and processing relationship to it.

Let us start with the latter relationship. (For the following see p. 196 ff.). Labour creates a universal form in that particular world by which the individual is surrounded or of which it is a part. Even at this early stage Hegel confronts labour with desire (*Begierde*), which is a relationship to the surroundings that is not yet completely human. (This concept of desire becomes even more significant as a contrast to labour in later texts.) Desire is in fact only directed at a specific and simple content. It negates itself as it destroys its object, e.g. in the moment when a natural object is consumed in order to achieve immediate satisfaction. It always starts from the beginning again. Labour, by contrast, changes the thing by giving it a universal form, based on universal ideas, and not the desires of the moment. Indeed these are often in this way prevented from being directly satisfied. The object of labour is nevertheless not destroyed, but preserved in a transformed state, as the surroundings of human society. Desire must then be distinguished from *needs*. Both Hegel and Marx perceived satisfaction of needs as a real historical process, in which the satisfied needs generate new and higher needs, and thus contribute towards the development of a superior type of human being.

In labour the object is recognized to a certain extent as an object with its own internal dynamics. This is necessary for the controlling and transforming of the object in a technical process. In the context of the application of tools, one part of this nature is played out against another in order to fulfil the purposes of man. In the processing of nature a preservation and acceptance of the external objectiveness is in fact presupposed. This view of technology is not new. It can easily be connected with Bacon's view of nature, knowledge and technology. But Hegel's interests are quite different to Bacon's. As Taylor points out, he emphasizes the fact that this acceptance of the internal dynamics of the outside world has repercussions on the subject and gives it a similar permanency to that of external nature itself.

In Hegel this idea can been linked to a further epistemological problem. For when labour stands for a relationship to the surroundings that surrounds the levels of desire, then we move beyond the idea that our relationship to our surroundings on its most original level is passive. Desire and sensing (in the traditional interpretation) share the property that they are both responses to given stimuli. In Hegel, the contours of a concept of active sensation are beginning to emerge.

Hegel discusses labour most extensively in the first of his main philosophical works, *Phänomenologie des Geistes* (The Phenomenology of Spirit) from 1807 (Hegel 1977) (referred to hereafter as *PhG.*). Labour is treated here in a different context to that of his earlier works. *PhG.* contains no stories of the gradual awakening of the mind from an animalistic state, nor any discussion of labour within a society with the modern divisions of labour and positive law. As regards this last perspective one could say that the place of labour has been pushed somewhat backwards; it is discussed in connection with the emergence of the self-conscious individual. The similarity to the Jena manuscript lies in the fact that labour is confronted with desire as a means of relating to the surroundings.

Labour is discussed in the single most famous paragraph of *PhG.*, which has the title "*Herr und Knecht*" (chap. 4 A, traditionally translated as Lordship and Bondage, but I prefer to use the terms master and slave). My reason for concentrating on this chapter here is the perspective of my book. I do not agree with the interpretations that would make it the decisive chapter in the structure of the book and that read into it a theory of capitalist society, interpretations which must be recognized as long since refuted (see, for example, Koyeve 1958). Indeed questions concerning the overall conception of *PhG.* reach beyond the bounds of my purposes. A decisive factor for my rejection of that form of Marxist weighting of this chapter, of which Koyeve is a representative, is the fact that the relationship between the master and his slave is a personal and thus pre-capitalist relationship of domination. The arguments for this point of view should be obvious from the following interpretation. (Of the works on this topic that I have found most illuminating, I would like to mention Gadamer 1973 and Armstrong-Kelley 1973).

Consciousness was for Hegel, as it was for Kant, grounded in self-consciousness, i.e. the relation of consciousness to *the self*. Hegel believed, however, in contrast to Kant, that it was in fact possible to give an account of the genesis of self-consciousness. (For the following see Hegel 1977, p. 111.) In Hegel's view a being can only be self-conscious by relating to another self-conscious being, not an inanimate object. A subject that only relates to its surroundings, i.e. external, organic nature, through desire can thus not be self-conscious. It cannot then be fully human, for in order to become human it must relate to another self-consciousness. The emergence of self-consciousness therefore implies an intersubjective relationship; the concept of an isolated self-consciousness is a contradiction in terms. Hegel's theory of the origins of self-consciousness thus constitutes the first draft of his theory of society which is presented later in the book in connection with the concept of *Geist*.

The relationship between master and slave is the result of the first meeting of two subjects that are striving to confirm their existence as self-consciousness. In this meeting the two self-conscious subjects join in a fight to the death. Both want to confirm their independence as self-conscious beings; the loser must submit to the other, give up his independence and become a slave in order to save his life. It is difficult to tell whether Hegel really justifies the idea that the first meeting between two self-conscious subjects *must* be a conflict or why he believed it to be so. It may be worth noting that in most significant political philosophies since Aristotle it has been accepted as legitimate to take prisoners of war as slaves. Even liberal theorists accepted the institution of slavery outside a political system.

Regardless of this, Hegel was not of the opinion that society is built on conflict or violence. On the contrary, he assumed that inter- and intrasubjective relationships had to be based on reciprocal relations. Since a being can only be self-conscious in a meeting with another being, a human society can only arise when both parties perceive and accept one another as humans. Hegel calls this kind of relationship recognition (*Anerkennung*), and recognition must therefore be a mutual affair. Only a human being that has already been recognized can give another the recognition he/she seeks. This is the point at which the master-slave relationship goes off track. The master wishes to confirm himself and is seeking recognition, but can only attain recognition from the slave, by means of the slave's relationship to himself. But the slave is treated like an object, he is not recognized by the master. The slave is therefore incapable of granting any recognition because he himself is denied it. In this way neither can the master attain recognition since he stands fast on his demand for one-sided recognition.

Jon Elster in particular has written in many different contexts about the master's self-contradictory project: Demanding and at the same time refusing to give recognition (see, in particular Elster 1979, chap 3.2). I can fully accept this interpretation which I have used as my starting point here, but my perspective in the following exposition is somewhat different.

The master forces the slave to labour for him, but does not labour himself. The master figure thus regresses to a kind of desirous, pre-social level, but with one major difference. This difference is not primarily implied by a change in the structure of the master figure, but rather in the nature with which he is confronted. What count as objects have changed for him. It is no longer a case of unmediated and obviously organic or inorganic nature. By means of the slave the master is now confronted with and relates to a processed nature, either the slave himself, or the nature which the slave has processed through his labours, a nature which we could say has been permeated with intentionality. The master's relationship to this transformed nature is not itself an expression of a human or self-conscious structure, but a form of animal relationship in a human or better social framework. The master figure thus remains a form of retardation. The master's relationship is essentially one of desire. As this relationship to concreteness is now a relationship to processed nature, it could well be called enjoyment.

> For the master, on the other hand, the *immediate* relation becomes through this mediation the sheer negation of the thing, or the enjoyment of it. What desire failed to achieve, he succeeds in doing, viz. to have done with the thing altogether, and to achieve satisfaction in the enjoyment of it. Desire failed to do this because of the thing's independence; but the master, who has interposed the slave between it and himself, takes to

himself only the dependent aspect of the thing and has the pure enjoyment of it. The aspect of its independence he leaves to the slave, who works on it (Hegel 1977, p. 116, [slightly altered translation]).

Despite the relatively positive assessment of enjoyment in relation to desire that Hegel provides here, it nevertheless remains that this is owing to the fact that enjoyment originates within a framework of social relations that have been created by the slave, not by the master himself. It is the slave who makes possible the master's enjoyment by negating the independent object via labour. Enjoyment is, no more than desire was, a relationship that allows the subject to relate to his surroundings and at the same time preserve himself as a self-conscious being.

It is worth noting that the master not only has a pleasure-based relationship to the nature negated by the slave, but also to the slave himself (or herself). This is of course not intended to represent a potential, new way for the master to realize a self-consciousness, for the slave himself only exists for him as a form of the negated nature which the master desires. But it entails that this relationship implies a broader concept of oppression than mere extortion of a surplus product. The slave is both subjected to forced labour and negated in the sense that his independence has been taken from him or her. The master is thus in a relationship both to a negated human nature and a negated external nature. The master's relationship to the slave will also be a relationship to a reified objectivity, for example in the form of sexual exploitation, and not only extortion of a surplus product. Furthermore there is no reason to assume that the master, who thus has a purely material relationship to both his physical and social surroundings, is capable of relating to himself as anything other than a thing. The sexual oppression, which in its vague outlines can perhaps already be recognized as sadism, can therefore reasonably also be expected to break into a form of sado-masochism. I am not going to expound on these themes here because they do not directly concern the epistemological significance of labour.[9]

It is this function of labour, its contribution to the formation (*Bildung*) of the working subject itself by means of the process of forming external nature, that Hegel accentuates in particular in *Ph.G.*, for this function is the decisive point in the slave's final superiority over the master. In order to

---

[9] I am assuming here throughout that the master-slave relationship is a concrete relationship between two subjects, and that we are situated within the realms of a set of problems regarding intersubjectivity. This is, in my view, the only interpretation which agrees with the structure of Ph.G. as a whole. A split in the individual consciousness can thus be deduced from this. For a contrasting interpretation which systematically regards the master-slave relationship as part of an epistemological problem linked to the isolated ego, see Julius Hartnack 1979, *Fra Kant til Hegel*. Copenhagen: Hartnack, J. 1979.

become human it is necessary also to be able to negate oneself to being and maintaining oneself in this negation. In connection with the role labour plays in man's formation of himself we can distinguish between two stages: First of all the working person negates his own spontaneity, which in turn postpones or excludes the immediate satisfaction of needs and himself becomes a source for a general forming of the world. This is only possible when universal and constant conditions in the external world are heeded. Thus the process of forming becomes as significant for the subject as for the object, for the subject is equally dependent on adapting to the internal dynamics of the object, as it is itself actively capable of creating a general form in nature. The labouring individual therefore has a moment of continuity and constancy, at the same time as he is reaching beyond himself and mastering his surroundings. In this way the slave conquers the master, labour conquers desire and pleasure and becomes an epistemological principle which also serves the establishment of societies.

> Through work, however, the slave becomes conscious of what he truly is. In the moment which corresponds to desire in the master's consciousness, it did seem that the aspect of unessential relation to the thing retained this independence. Desire has reserved to itself the pure negating of the object and thereby its unalloyed feeling of self. But that is the reason why this satisfaction is itself only a fleeting one, for it lacks the side of objectivity and permanence. Work, on the other hand, is desire held in check, fleetingness staved off; in other words, work forms and shapes the thing (p. 118, slightly altered translation).

But this concept of the formative function of labour is only one part of man's development into man and cannot replace intersubjective relations, which are found in the concept of reciprocal recognition. Before I go any further, I would just like to point out that for Hegel labour is only a formative function when it appears in a configuration with absolute fear and being forced to perform a service. But regardless of whether labour is an independent pattern in Hegel's philosophy, labour is the decisive factor for creation *within* this configuration. Fear is mediated by means of subjection to forced labour. Without the absolute, experienced fear of death, labour becomes devoid of the totally negative aspect where the subject refinds himself, but without labour, fear becomes "introspective and mute" (p. 119). It is obvious why Hegel had to claim this. He needed a category which can explain the fact that the slave generally reduces himself to being, otherwise labour would not be the response to a general situation. Perhaps the best interpretation would be that force was a historical precondition for a developed labour process.

It is however more dubious to claim that the fear of death alone, as a deeply felt shock, would be able to achieve the same character of a constantly forming element as labour has done. Although Hegel perceives the three factors labour, fear and force as necessary, it is unreasonable to ascribe this particular interpretation to him. Fear is the necessary incitement for creative activity, but labour is the necessary moment in the actual formation. That "wisdom" which only has its "beginnings" in fear "comes to itself" via labour (p. 118). As far as forced labour is concerned, it is most reasonable to interpret it as a form of discipline. (I am assuming here that in this part of his text Hegel is not dealing with capitalistic production where the relations between the masters and the workers are impersonal, and discipline is primarily a euphemism for forcing people to submit to the purely social conditions inherent in the process of production.)

In the following I am going to attempt to summarize what appear to be the basic characteristics in a theory of the epistemological function of labour and tools and of active sensation or, active intuition, without following Hegel's text too closely. In connection with the latter concept let us dispose of one fundamental preconception. I disagree with the common conception that prior to the upstart of labour operations, one must have already sensed certain qualitative and quantitative properties of the world, of objects in the world, and have a concept, no matter how primitive, of the internal dynamics of these things and the possibilities for satisfaction of needs in various natural environments. The very observation of labour in an epistemological context ought to teach us to discard the idea that perception is basically a passive relationship to the outside world, or, to put it differently, that affection comes first, then action.

I referred to the historical development that led to this point of view at the beginning of this chapter. Epistemological arguments for this kind of view can be found in the works of writers like M. A. Ayers and John Gibson. Ayers (Ayers 1970) demonstrates that even traditional theories about the passivity of sensation (traditional in comparison to theories about labour as active sensation) themselves presuppose the possibility for certain forms of activity and actually base their arguments on them. Activity in this sense must be understood as the possibility of executing bodily movements, as use of the body's motor apparatus.

It is this form of activity that is significant in Gibson's discussions (Gibson 1966). Bodily movements are of utmost importance for sensible intuition, because it is through them that we receive information from the sense organs. First of all we must assume that we place ourselves, our sense organs, in such a position that we can be affected, and secondly that our being affected occurs in connection with movements of the body, like, for example, when we run our fingers over an object. Beyond this it is not only the movements of the parts of our bodies, but also their relative positions that play a part in perception. The fact that we can actually feel something through the end of a stick, e.g. if the object we are touching with the stick is hard or soft, is due to the interrelations between the organs (Gibson p. 113). Gibson also claims that there is a difference at the physiological level between information obtained through actions and that received via passive stimulation (p. 39). Similarly Dux (Dux 1982) claims that motor functions of the body are a precondition for experiencing objects. To claim that information from a passive influence must come before and is itself a precondition for action is therefore incorrect because actions in themselves are an essential source of *obtaining* sense information at first. (For the entire discussion see Krogh 1985, chapters 4 to 6.) This problem has also been dealt with in the pragmatic tradition in modern philosophy. I have chosen in this context to limit myself to modern references that shed light on the question in relation to the philosophical debate at Hegel's time.

I shall divide the further discussion of labour into four short subsections. They are A) the object, B) the subject, C) the means or instrument of labour (i.e. the tool or machine) and D) the configuration constituted by all these moments. The division of the labour process into these moments is taken from Marx's outline (Marx 1965, p. 177). As a whole their purpose is to provide an outline of the concept of active sensation. It is therefore impossible to avoid a number of cross references within the representation. I have chosen to make them explicit rather than to ignore them.

A) Labour can be regarded as a means of collecting experiences about the things around us. Through labour we acquire knowledge about their natural properties, i.e. their internal dynamics and distinctive qualities. We learn about their regular, law-like reactions to human intervention and influence from other natural objects. Because of labour our need to find out more about these things also grows, in order for us to be able to predict their future states, "more" here in the sense of more than is readily available to the disinterested observer. In this way we can regard labour as a source of empirical insight into nature and of a means of establishing systems for the classification of natural phenomena. These systems will, needless to say, be relative to the historical level that the labour processes of the various societies have achieved. Even Locke, writing within the empiricist tradition, claimed that pragmatic requirements to communication are fundamental to our empirical systems of classification which do not reflect the real divisions of the world. Marx underlines the role of the establishment of concepts in securing self-preservation (see Locke 1961, III, vi, 30 and 36, and Marx 1966).

Even if we accept that it is artificial to let sensory experience start on a purely receptive level, there are nevertheless many objections we could raise to ascribing such a prominent epistemological role to labour, as only one of many human activities. We could claim that the kind of bodily movements we are dealing with here are not restricted to labour situations; they appear just as frequently in, for example, children's games. It is therefore possible to say that it is untenable to put forward the idea that the relevance of these kinds of activities allows *labour* to occupy such a fundamental position. There are however reasons for attaching special importance to the accumulation of knowledge via labour. It seems plausible to me that it is only in connection with labour, or at least primarily there, that we are seriously confronted with the internal dynamics of nature and natural objects, and it is only there, or at least primarily there, that man is forced to interfere with these dynamics systematically by presupposing them and exploiting them. This theory is, however, only significant in an epistemological context. A word of warning about the risks involved in an attempt to use these kinds of theories about labour as general theories of socialization: They cannot serve as models for complete theories of the establishment of societies and the process of civilization, only as parts of them. They only indicate a single element in this process, as I hope the discussion of Gehlen made clear.

The perspective I have laid out so far must however be broadened by emphasizing the fact that the objects of labour form part of an already processed nature. The theme I have touched upon here is of course closely related to the topic known as the inertia of materiality. As this term is used in phenomenologically inspired Marxism, it is also often used to designate the meeting with previous intentionality, which is already inherent in the social materiality, as I described in the Introduction. This indicates a key word for a more essential function of labour than that as a relationship *to* objects. Labour does not only give us knowledge about consciousness-independent objects, subject to their own internal dynamics. The activities which comprise the labour process also alter the object. They also create a new world, a world of artefacts instead of objects.

The distinction between artefacts, or products, and objects is certainly both fluid and relative to various perspectives: Any product *can* of course also be regarded as an object. This kind of perspective forms the foundation of the scientific means of observation, and the phenomenological perspective is not intended to compete with that. My point is that here we are studying the surroundings of society as they appear in the light of the metabolism of society, not on the basis of physics. The essence of the epistemological perspective is then: We, human beings as a specific subject, are always surrounded by a world that has arisen as a result of the human intentionality invested in it. In this world we are always "on the inside", so to speak, surrounded by an objectivity with which we are well acquainted because we have created it. This point is shown to greater advantage when we compare this constitution of the world with the constitution of the subject.

B) If we start off by treating the subject in isolation we are able to recognize the same pattern. First of all it is possible to perceive labour as a source of knowledge for the subject. This corresponds, for the subject, to what I said above about labour as epistemological activities. Moreover the main point even here is also the transformative aspect of labour, which in connection with the subject we can call its formative function. Below I am going to present three aspects of the way in which labour can produce a new subject, in the same way that it can also produce new surroundings.

1. Labour can be said to generate a new physical structure, or to generate a real, i.e. species-specific human structure, when direct satisfaction of needs on a systematic level is made subordinate to a conquest of the internal dynamics of nature, which again presupposes capitulation to these. Postponement, waiting and planning, in other words negation of the immediate, are the key words for this type of formation, which I have discussed in greater depth above, in both this and the previous chapter.

2. Labour can be perceived as a moment in the development of purely cognitive skills. This applies in particular to the ability to use general, idealized conceptions, especially at such an early stage that we cannot say with certainty whether man was capable of giving verbal expression to these skills. The ability to create two or more identical products, i.e. production based on a prototype which need not to be directly present, presupposes the ability to imagine and apply a general concept. This ability was already present in early humans (cf. Oakley 1959 and Beck 1980, chap. 7). Again it is perfectly reasonable to view labour as a contributive factor in the routinization of these skills, but not necessarily as their catalyst.

3. In my context a different feature is of even greater importance. Comprehension of the epistemological character of labour does not only pave the way for an understanding of the significance of activities for the acquisition of sensory information. It also admits an understanding of the *historically variable nature of our very perceptual apparatus*. The most important epistemological function of labour in relation to the subject is precisely the way in which it preforms and transforms our way of perceiving the world. A classic account of this aspect of a materialistic epistemology is given in Max Horkheimer's classic study from 1937 (Horkheimer 1937). If we add this feature, then it becomes reasonable from yet another perspective to say that there is no pure nature as long as we apprehend our surroundings in a social, as opposed to scientific perspective. Neither is our way of apprehending

each individual object, even if we ignore any forming in a labour process, independent of the social forming of our sense apparatus through a labour process. It is then reasonable to say that points 1 and 2 are most important in the first stages of the process of civilization, such as that which we find in early humans. Their decisive function occurs at a stage prior to Hegel's level of recognition, even though all societies will have to reproduce themselves by means of a metabolism with nature. They are not constitutive for the specific in the emergence of human societies in the same way. Point 3, which is directly linked to the idea of materiality, will be relevant on all levels of society, an aspect of all periods of society. An example of an application of these theories in a context which combines aesthetic practice with a high level of theoretical abstraction is Walter Benjamin's argument that film teaches the individuals in modern society a way of perceiving their surroundings that is on a level with modern reality itself (see Benjamin 1936).

Another interesting follow-up of these thoughts is found in André Leroi-Gourhan. He has developed a theory about what he calls mechanical rhythms or chains of operations (Leroi-Gourhan 1980). This rhythm is engrained in the subject through learning and adaptation to the surroundings. It is a relationship to one's surroundings that affects both the motor functions and the consciousness in an attempt to systematically use the laws of nature for one's own purposes. It is rooted in neither a biological determination nor pure spontaneity. It lies somewhere between the biological rhythms of digestion, breathing and sexuality and the innovative use of the motor *and* the cognitive functions in activities such as dancing, sport and artistic creation. This appears to be a fruitful way of describing how biologically given reactions are captured in social patterns of a more complex character than they themselves can determine and thus contribute to rendering them possible.

At the same time this demonstrates a division amongst those theories which are linked to the concept of social materiality. Hegel, Marx and Sartre agree with Leroi-Gourhan, since they show how technology creates mental and physical routines which are engrained in the subject and are not spontaneous or original. Merleau-Ponty and Heidegger are more diffuse here and seem to lean more towards a view of technological actions as original output. Heidegger locates, for example, his existential *at hand* above biological responses and under intended action (Heidegger 1927, § 15). But his descriptions do not allow for the comprehension of the fact that the use of tools he describes, no matter where it is placed in the user's repertoire of actions, is acquired and determined by the technology of one specific period. From the point of view of the history of technology, the distinction between reflexive and pre-reflexive action is pretty irrelevant.

C) Through the tool the two creative processes I have described above are mediated in both directions. On the one hand the tool exists as an especially privileged part of what I have called the created world. It is the most objective expression of human intentionality, the external condition for realization of our purposes. Seen in this way it is more than a mere extension of the body and of the ability to perform actions. It exists as a universalization of our purposes. Furthermore the tool is created to be used several times, and it is an example of a model, an ideal object, which has been the basis of many artefacts. In this way it channels and generalizes the intentionality in relation to external things and appears to me and others as an objectivation of this intentionality.

On the other hand it is also a canalization of the laws of nature in relation to the worker. Engrained in the tool is not only my intentionality in a generalized and objectivated form, but also the knowledge about and exploitation of the regularity of nature, which are necessary for the realization of our intentions. Thus through use of the tool or instrument in the widest sense, not only the worker's formation of another object is mediated, but at the same time also the worker's formation of himself. In this way the tool is an expression of the field of improvization, renewal and attainment of goals in connection with the transforming of the objective reality, but also of the fetteredness to and dependency on these same surroundings.

D) Let us finally summarize these theories about the epistemological function of labour, the theory of active sensation, by regarding all these moments in context. The essence of the matter is then that labour within this tradition, considered as the situation where the subject produces the object and itself in a constant process, constitutes a constellation between subject and object. In labour as a constellation between subject and object the two poles are always mediated in advance through the previously performed labour, and this mediation both produces and is borne by what I have called materiality.

This function of materiality becomes clearer when we also draw in the perspective of time. The model I have developed here can be used both synchronically and diachronically. What I have said so far probably has a smack of the synchronic perspective. The diachronic perspective will of course become significant as we approach more concrete historical descriptions. As a part of the social materiality the tool, or the means of production, will combine past, present and future. It links us to our previous selections (under restricted conditions) and contributes to determining the body of possibilities which are available in the near and not-so-near future. Seen in this way materiality is a medium for incorporating knowledge about the objectivity of the succession of time.

# 4 Karl Marx

In order to provide an overall impression of the theory of active intuition and social materiality, and at the same time indicate the limits of a conception which perceives history according to the pattern of a collective subject's products, it might be fitting to start this chapter off with a quote from Marx in his polemic against Feuerbach:

> For instance, the important question of the relation of man to nature (Bruno [Bauer] goes so far as to speak of "the antithesis in nature and history" (p. 110), as though these were separate "things" and crumbles of itself when we understand that the celebrated "unity of man with nature" has always existed in industry and has existed in various forms in every epoch according to the lesser or greater development of industry...Feuerbach sees only factories and machines, where a hundred years ago only spinning-wheels and weaving-looms were to be seen, or in the Campagna of Rome he finds only pasture lands and swamps, where in the time of Augustus he would have found nothing but the vineyards and villas of Roman capitalists...So much is this activity, this unceasing sensible labour and creation, this production, the basis of the whole sensible world as it now exists, that, were it interrupted only for a year, Feuerbach would not only find an enormous change in the natural world, but would very soon find that the whole world of men and his own perceptive faculty, nay his own existence, were missing...He does not see how the sensible world around him is, not a thing given direct from all eternity, remaining ever the same, but the product of industry and of the state of society; and, indeed, in the sense that it is a historical product, the result of the activity of a whole succession of generations, each standing on the shoulders of the preceding one, developing its industry and its intercourse, modifying its social system according to the changed needs. Even the objects of the simplest "sensuous certainty" are only given him through social development, industry and commercial intercourse. The cherry-tree, like almost all fruit-trees, was, as is well known, only a few centuries ago transplanted by *commerce* into our zone, and therefore only *by* this action of a definite society in a definite age it has become "sensuous certainty" for Feuerbach (Marx 1969a, p. 43 ff. English translation taken from *The German Ideology*, C. J. Arthur (ed.) New York 1970, pp. 62–64).

So here we have in a nutshell the whole idea of the collective subject which processes external nature, a theory that we today associate first and foremost with Marx. My discussion of Marx in this chapter however is on a different plane; it is linked to Marx's account of the role of technology in the capitalist mode of production, paying particular attention to the machine and the factory system.

I would like, by way of an introduction, to air my views on two exegetic problems which arise in connection with Marx's view of technology. The first is the question of whether Marx was a so-called technological determinist. I am defining the concept of technological determinism here in the way I deem to be the most common, namely as a theory that claims 1) that technological development determines all other social relations, and 2) that the development of technology itself is given independently of all other circumstances (Feenberg 1991, p. 130). The quotation that is most frequently cited as evidence for attributing this kind of interpretation to Marx is, of course, those famous lines from "The Misery of Philosophy" where he says that:

> The handmill gives (*ergiebt*) you society with the feudal lord, the steammill, society with the industrial capitalist (Marx 1964, p. 130).

Let me offer a contextual argument against interpreting this as technological determinism: It is not unreasonable, but neither is it necessary to interpret this "ergiebt" as expressing a theory of an unequivocal determination. An essential element in this work is a polemic against what Marx claimed was Proudhon's tendency to try to understand circumstances particular to specific historical periods from the point of view of universal categories. It is therefore in keeping with this tendency that Marx in the same book says:

> Machines are no more an economic category than the ox before the plough, they are only a productive force. The modern factory, based upon the use of machinery, is a social relation of production, an economic category (p. 149).

The most reasonable interpretation of this quotation is undoubtedly that Marx is claiming that the social role of technological systems can only be understood within the framework of the economic conditions which are prevalent at the time in question. The quotation refers to the limited usefulness of a point of departure that abstracts technologies from this kind of context, and is therefore actually an expression of the opposite of technological determinism. The desire to attribute a technological deterministic interpretation to Marx merely on the grounds of this famous quotation is most certainly a case of poor textual exegesis.

Regardless of the way in which one chooses to interpret Marx's *The Misery of Philosophy*, one must of course concentrate on *"Das Kapital"*

(Marx 1965), and here Marx explicitly states that the capitalist use of the means of production, itself an economically determined condition, determines the technological development.

> In the capitalistic process of production the process of labour appears only as a means, the... process of production of surplus value (*Verwertungsprozess oder Produktion von Mehrwert*) as the aim (Marx 1969b p. 29).[10]

(For comments on these concepts see below). I would therefore join Donald MacKenzie (MacKenzie 1984, p. 473 ff.) in stating that this is a decisive argument against interpreting Marx as a technological determinist in the above-mentioned sense. Lurking behind the rejection of this interpretation is of course the opinion that technological determinism is an untenable theory. Jon Elster, by contrast, actually characterizes Marx's theory as technological determinism, but he seems to be using the concept in a different way from me. If Marx's technological determinism is to be rooted in the fact that he claimed that the rise and fall of successive property regimes are explained by their tendency to promote or fetter technological change (Elster 1986, p. 105), then technology in itself is not perceived as both all-determining and exogenously given. The best one can then say is that there is always a tendency to rationalize labour, that is to produce the same amount of product with less labour input. As far as I can see, nor does George Cohen go further than to advocate such a theory (Cohen 1968). This is not technological determinism in the sense in which I am using the term.

I am not going to spend much time considering the second problem either. It deals with the interpretation of the famous text from 1859 where Marx seems to introduce a general and "official", albeit very compressed, theory of history, or more precisely, theory of historical change. This is due to the fact that a satisfactory discussion would also have to go into areas which are far removed from the main points in Marx's theory of the development of technology under the capitalist mode of production. Regardless, it is most unclear whether what Marx claims here or appears to claim, conforms with

---

[10] This is not in fact strictly speaking a quotation from *Das Kapital*, but from a work which is known as *Resultate des umittelbaren Produktionsprozesses* (Marx 1969b). This is however a manuscript that Marx had planned as the final chapter in the first volume of *Das Kapital*, but which he edited out during the final review. So there can hardly be any doubt that this manuscript belongs to the same stage of development as the main work. Nowadays it is printed as an appendix to new editions of *Das Kapital*. As we shall see, it contains some of Marx's most important concepts for the definition of the relationship between technological development and wage labour. It has received very little attention, which has resulted in major flaws in most discussions of Marx as a historian of technology. A praiseworthy exception is MacKenzie.

the deeper and historically more specific analyses of the production equipment of the capital in *Das Kapital*. No reader of either *The Communist Manifesto* or *Das Kapital*, with their descriptions of the enormous changes which the bourgeois mode of production continuously entails, can help but wonder whether it is not in fact the relations of production that have a progressive effect on the means of production. This of course means it becomes doubtful whether Marx actually put forward a consistent general theory of the development of history. For my purposes however, which are concentrated on the technology shaped by capitalism, further comment on this text is unnecessary.

We do however need some of the basic concepts from Marx's general theory of history in this context. Marx used the term *productive forces* to mean all the aspects that promote the ends-means rationality of labour, which ease our contact with nature and enable us to produce the same amount of product with a smaller input of manpower than previously. If we here include labour power, its education and organization in the concept of productive forces, as Marx did in places (Marx 1965, p. 40), then we have another piece of evidence against interpreting Marx as a technological determinist. The *relations of production* are by contrast to be understood as the social relations of production, especially the legal ones.

The *process of labour* refers to the superhistorical or universal elements, those linked to labour in all ages. The *capitalist process of production* is the shaping of the process of labour in a period with free wage labour and the right to private ownership of the means of production. Shaping in this context is obviously to be understood as more than a mere modification of the original terms.

The main concept in Marx's economics, in keeping with the entire tradition from Adam Smith and David Ricardo, is *value*. The *value* of an individual product is the number of *working hours* (not individually, but calculated on the basis of a social average) invested in the production of it. The *use value* of the product consists of the useful, physical properties of the product. If we now assume that all products are commodities we are able to answer two questions. We can now see how a universal increase in wealth is possible, why the economy is not like life itself, in the sense that you actually get more out of it than you put into it. The theory of value also has pretensions of being able to answer the question of what it is that determines the relative prices of various goods. If it were the balance between supply and demand, the *relative prices*, expressed in money, it would, in a suitably simplified model, correspond to the number of working hours invested in the various products. This is not however the time or the place to go into Marx's theory of value. In short we can say that within a dialectical version of the theory of value a distinction must be drawn between the surface and the deep structure of society. Value appears on the surface as money. Questions as to how this so-called transformation is supposed to take place, and whether this concept is at all meaningful, are both controversial themes within the theory of value itself and have triggered off weighty criticism of the theory of value as a whole.

In a capitalist society we must furthermore assume that *working power* too is a commodity which can be bought and sold, that wage labour exists. The price of labour power is the *wage*. This commodity is the only one that can actually produce value and it can produce more value than its owner, the worker, receives in remuneration, i.e. wages. The surplus of value devolves on the owners of the means of production. This part of the value product is called *surplus value* and the capitalists' appropriation of the surplus value is *exploitation*. (The *profit* is the surplus value expressed in money.) It is appropriate to mention here that the entire theory of value is controversial today. We do not however have to swallow it whole; it is possible to work out a concept of profit and capitalist exploitation without accepting or presupposing the theory of value. It makes no difference to our description whether we say that production is determined by the need for increased surplus value or profit. We can now see the implications of the quotation cited above. Under capitalism the universal moments of labour are turned into means for the extraction of profit and extraction of profit to an exaggeratedly increased degree.

It is also important to underline that *capital* in this context does not primarily mean the money that has been invested, but also what it has been invested *in*, the production equipment i.e. machinery and buildings, raw materials and semi-manufactured goods, basically all the materials which the wage-earner uses in his labour. Capital which has been invested in production equipment, raw materials etc. is called *fixed* capital. The point of departure of Marx's entire theory of technology is that technology exists as fixed capital and that all technological change must be regarded in relation to the profit motive (and of course the resistance of the working class, the class struggle).

An account of Marx's view of the relationship between technology and technological change and the capitalist relations of productions can be formulated on two levels, referring to A) a diachronic and B) a synchronic point of departure. The synchronic approach consists of starting with a universal and superhistorical description of the labour process and seeing how the moments in it are mediated in the period of capitalism. The diachronic approach by contrast consists of studying how the capitalist relations of production alter the contemporary technological and organizational relations of production, which they themselves have inherited from earlier periods, and how they change during the various phases of capitalism.

A) The definitions of the material aspects of the capital have their point of departure in the general definitions of the labour process. Marx describes them thus:

> The elementary factors of the labour process are 1), the personal activity of man, i.e. the work itself, 2), the subject of that work and 3), its instruments (Marx 1965, p. 178).

> Through the nature of the process of labour the means of production firstly divide themselves into means and into objects of labour. These are forms of determination of use value which arise from the very nature of the process of production (Marx 1969b, p. 7).

But when the labour process becomes a capitalist process of production, i.e. subordinate to the fundamental objective of producing for profit and increased profit, it appears that the threefold division that was in a state of equilibrium is pushed aside in favour of another distinction. In addition to the division on the side of the object between means of labour and object comes

> ...secondly, in a conceptual division which has its source in the process of labour between the objective conditions of labour (the means of production) and the subjective conditions, the teleologically active ability of labour, i.e. the labour itself (p. 9).

Here we can see the arrangement of the purely physical components of the labour process (we can interpret labour here as a physiological process) as they appear in a production under the capitalist mode of production. The essential distinction within the capitalist production process is therefore between labour and its material conditions. This distinction is

> ...a further determination of the form of capital as far as it appears *sub specie aeternitatis* within the immediate process of production (p. 8).

Through his extensive production Marx developed various different terminologies to describe this relationship. In *Grundrisse* (Marx 1953) he speaks mostly of the relationship between *living* and *dead* labour. Labour as an on-going process is in opposition to and is subordinate to its own finished products. The output is captured by and subordinated to its own objectivation. We can regard this terminology as an echo of the more anthropologically coloured interpretations of labour in Marx's early writings, which still bear the marks of the expressionistic model we found in Hegel. In *Das Kapital* this terminology is replaced by a more neutral distinction between the subjective and the objective conditions of labour or simply by the economic terms labour and capital.

This development constitutes a parallel to the development in Marx's theories about alienation. In Marx's early work we come across a theory of alienation primarily as alienation from oneself. The theory of the distinction between subjective and objective conditions in the production process is also a theory of alienation, but here Marx describes first and foremost the relationship between man and his *surroundings* as alienated. The idea of the relationship between living and dead labour is in an intermediate position; in the idea of the surroundings as a coagulation of our processual being, the idea of man's alienation from himself is still present in the background.

There is no obvious explanation of exactly what function the reference to a superhistorical theory of the labour process has in Marx or of the manner in which he imagined that the various forms of processes in the different periods are in fact mediations of the basic model. First of all, it is not possible to overlook the fact that this model of labour in itself is Aristotelian, and most probably is based on a craft-based model of labour, rather than on a model adapted to modern relations of production. I am not thinking here first and foremost of the fact that these kinds of craft-based theories often have aesthetic overtones. It is the actual threefold division which becomes problematic when seen in the light of modern relations of production. If we take the example of the original chemical industry, which was just starting to develop in Marx's time, it is at this stage already problematic to claim that the product exists as a processed version of an object that was present at the outset of the labour process, and it is all in all highly difficult to talk of any kind of distinction between the means of labour and the object. It thus appears that what I have called the super historical account of the labour process is not adequate as a smaller common multiple of the purely physical development of the labour process in all periods.

One can further imagine it as a *"Ding an sich"* theory; the labour process has a fundamental structure which we know only *through* the various historical mediations of it. Neither is this interpretation, however, particularly appropriate, for Marx indicates precisely what the universal moments in the labour process should be.

There are of course other possible interpretations. We could perceive the general theory as a description of the labour process at the earliest, most primitive stage. But Marx seldom made the mistake of taking what was *early* to be the *essence* of the matter, of which the later periods were mere mediations. And even if we did ascribe to Marx the theory that there is a fundamental tendency in all historical periods to rationalize the means of production, he did not perceive successive *economic* periods as stages in a cumulative process. Antiquity, feudalism and capitalism all build on their own characteristic and very different principles. (Marx's attempt to summarize the whole of history on the basis of categories in Hegel's dialectical logic is too general to be interesting, and positive tendencies to deduce the historical development from abstract principles are at variance with his basic beliefs to such a great degree that we should be wary under any circumstances of attributing such tendencies to him.)

These objections only arise if we choose to understand the theory of universal moments in the labour process as a *framework* for all historical periods. It is not easy to recognize what possible function a theory of this kind of framework could have if it is nevertheless impossible to deduce the principles for the economic structures of the various periods from it. In Sartre, whom, in contrast to Marx, I want to interpret as a deliberately transcendental philosopher, there is however room for such a framework of principles, the aim of which is to render history comprehensible. (See the next chapter.)

I deem it most profitable today to ascribe to this theory a double function. On the one hand it can be interpreted as an expression of a naturalistic element in Marx's thinking; he wants to accentuate the fact that all societies in their interaction with nature, what Marx called metabolism *(Stoffwechsel)*, are confronted with an internal dynamic in the natural materials that are processed and in the natural processes that are exploited (cf. Schmidt 1963). This view is still relevant with regard to theories like "the social construction of technology" which appear to perceive artefacts as pure projections of cultural relations. But its field of relevance is limited today, as we have already mentioned, since many modern technologies, even if they are included in what we can call productive activity, cannot really be conceived of as a step in the processing of nature.

On the other hand we can successfully ascribe to this theory a heuristic function. It equips us with a set of elements that allows us to grasp, from a bird's-eye point of view or, as I said above, in a synchronic perspective, the mutual differences in the relationship between the various elements in the various periods. The application of this kind of approach to capitalism will result in *the* specific arrangement of the three elements of the labour process, seen in relation to the bifurcation of the labour process into a subjective and an objective side, that is typical for the capitalist process of production.

It is then obvious that this kind of perspective is *"gefundenes Fressen"* for a structuralist approach, which in turn is incapable of grasping the emergence and alteration of a system. I am following Etienne Balibar in this context, who in my opinion gives the best account of this theme, albeit within the restrictive perspective of structuralism. Balibar interprets the industrial revolution as a change in the relationship between the elements in the labour process.

The industrial revolution (the transition from manufacture to modern industry) can be completely defined by the help of these concepts, as *the transformation of the relationship* which followed from the replacement of the means of labour. ...it could be represented as the succession of two 'material forms of existence' (Balibar & Althusser 1970, p. 242).

This can be illustrated as follows (I have simplified and changed some of Balibar's terminology):[11]

| *manufacture* | *large-scale industry* |
|---|---|
| labour | |
| | > production of tools |
| means of labour | |
| | > mechanical industry |
| object of labour | |

It is not clear at which level, in relation to which type of entities, Balibar believes he is able to trace this kind of change. Perhaps it is most reasonable to say that he provides an analysis on a techno-sociological level. But this perspective is too narrow. The diagram above illustrates a fundamental distinction between the subjective and the objective elements of labour. The decisive point is that this distinction appears on *two* different levels: The legal level (or level of the relationships of ownership) *and* the technological level (or level of the means of production). In a system of wage labour the worker would be judicially free and meet the capitalist in the labour power market as a seller of a particular commodity: His own labour power. (This distinguishes societies with wage labour from all those forms of society where the worker is bound by law and subordinate to another social class, such as in slavery and other forms of exploitation in feudal societies.) The means of production and the end product belong to the capitalist; the means of production exist as fixed capital. The producers are thus *judicially* separated from the means of production.

We discover the same configuration on the technological level: The labour power, what Marx called the subjective aspect of labour, is separated from and is in opposition to the material or objective aspects of the produc-

---

[11] The term manufacture does not actually designate a precapitalist, but rather an early capitalist form of production. Since however it is based on a premechanical craft-based technology, a comparison with manufacture can serve to accentuate the specific elements in capitalist large-scale industry.

tion process. It is this *merging of a judicial and a technological separation of the worker from the means of production* which for Marx constitutes the essence of mechanical technology under capitalist conditions. By means of this double separation the new relationships of ownership are in this way given an adequate technological expression here.

The very idea of such a definite, necessary, technological level is quite suspect in this day and age. Marx seems to overlook the national, cultural and epochal adaptation that even the capitalist production equipment has to undergo.[12] Marx would probably reply that, irrespective of local development, capitalism was essentially linked to machine technology and large-scale industry. Neither does Marx discuss contrafactual hypotheses such as whether a capitalist system is at all possible (especially a capitalism outside agriculture) without machine technology and the factory system. Nor does he deal with the question of whether there could have been alternative routes to the development of machine technology other than that which occurred at the end of the eighteenth century, e.g. if other sources of energy, which we now know about had been used extensively instead of coal and steam. His main thesis seems to be that a capitalist system, i.e. a system of wage labour and the right to private ownership of the means of production, contains a dynamic that must result in transformations of production techniques that fulfil certain requirements. The capitalists actually discovered this kind of technology via the development of machine technology.

Based on my observations in the Introduction, however, I still wish to retain my reservations about the idea that capitalism is necessarily linked to one specific technology. Below I am reading Marx's account as a description of the necessary conditions for the actual development in Western Europe, not for every possible form of capitalism.

B) We can now move on to look at Marx's diachronic approach to the problem, the representation of the development of the labour process under capitalism towards the factory system, as it was known in his times. Marx claims that the domination of capital over labour historically assumes two successive forms, which are related to their own technological levels. The organization of wage labour, not only judicially but also technologically, develops under the capitalist mode of production from an earlier, more primitive technological stage. The relations of capital we know today exist on the basis of a specific technology which did not appear overnight.

Marx's most important concept for describing how a capitalist production system transforms technology from an early stage is the distinction between *formal* and *real* subsumption under the capital. These key concepts

---

[12] For a clear formulation of this view, see Sejersted, F. 1990, *Er det mulig å styre utviklingen?* Paper no. 11. TMV-centre.

have been seriously overlooked in literature on Marx hitherto, despite the fact that the distinction between formal and real subsumption is fundamental in *Resultate*. This inability to recognize the significance of these concepts can perhaps be blamed on the fact that they are not used in those parts of *Das Kapital* which contain Marx's main account of the historical change of the production process under capitalism, namely chapters 13 to 15, but first appear in chapter 16, where Marx regards the organization of the labour process from the point of view of the maximization of profit, not of the history of technology.

In chapter 16 he introduces the concepts of real and formal subsumption in connection with the distinction between absolute and relative surplus value. Surplus value is, of course, as mentioned above, the amount of value created by the worker over and above the amount of value he or she receives in the form of pay. If this kind of surplus value can only be achieved by continuously lengthening the working day beyond the number of hours spent creating the aforementioned amount of value, then we are dealing with absolute surplus value. If, however, greater surplus value is created by means of technological or organizational changes, then we are dealing with relative surplus value. Technological change is therefore primarily a means to increased extraction of relative surplus value.

Marx writes:

> The production of absolute surplus value turns exclusively upon the length of the working day. The production of relative surplus value revolutionizes completely the technical process of labour and the composition of society. It therefore presupposes a specific mode, the capitalist mode of production...In the course of this development the formal subsumption is replaced by the real subsumption of labour to capital (Slightly altered translation Marx 1965, p. 510).

Marx uses the concept of a distinction between formal and real subsumption in two ways. As we gather from this quotation, the expression formal subsumption, is the common name for the relationship of labour to capital, i.e. an expression of the judicial relationship of wage labour to the means of production. Formal subsumption becomes real subsumption at the point where wage labour and the entire capital relation that is based on it reach an adequate level — as far as technology is concerned. Then the distinction between worker and means of production is implemented on both the judicial and the technological level. At the same time, however, formal and real subsumption are, in Marx, terms for two consecutive periods, two stages in the technological development of capitalism. The general characteristics of capitalism appear in the first period as its characteristic properties. All exploitation of alien labour is:

> ...the formal subsumption of labour under capital. This is the universal form of all capitalist processes of production, but it is also a specific form parallel to the developed specific capitalist mode of production (Marx 1969b, p. 46).

At the transition from formal to real subsumption, understood in the last sense as the transition between two historical phases, a shift in technology occurs, a decisive technological development which is a necessary condition for capitalism in the form we know it.

Formal subsumption as a separate historical period is not itself linked to any transformation of the production equipment. It "differentiates itself only formally" from the previous mode of production, out of which it directly grows (p. 51). Subsumption is only expressed here through "the cash nexus": The financial affairs that connect the employer and the worker. But as a further development of formal subsumption as a specific period (it also survives in branches of developed capitalism which have not yet undergone a technological revolution) Marx predicts the emergence of:

> ...a technological and in other ways specific mode of production, which alters the real nature of the process of labour abd its concrete conditions — the capitalist mode of production (Marx 1969b, p. 60).

The decisive element in this real subsumption is the modern factory system, that is to say the application of machinery, or rather machine technology, on a large scale.

We need to take a closer look at how Marx envisaged what he termed the adequate link (adequate from the point of view of capitalism) between machine technology and capitalism. I will return to Marx's definition of the terms machine technology and capitalism and his interpretation of the technological separation from the worker later, as these terms can best be illuminated by means of the description of the development towards real subsumption. I would just like to make it clear here that the concept of machine technology must include the new forms of sources of energy that machine technology required, the steam engine. We can, however, already begin to make out what this theory implies with regard to the relationship between machine technology and capitalist society. But first some more general historical reflections.

It is at any rate certain that when technology assumes the form of fixed capital, fundamental changes ensue on both the cultural and the social levels. Capitalism is without a doubt the first form of society that not only economically, but also culturally, psychologically and ideologically, is not only positive towards, but in fact requires and presupposes institutionalized technological change. In this way it differentiates itself from the technological development in the Middle Ages as described by Lynn White (White 1963). All cultural barriers against seeking such a change or applying innovations, not to mention voluntary deconstruction of the established technology, disappear, at least in the spheres involved in production. The numerous cases of reservation with regard to innovations, such as the regulation of speed limits for cars and trains prior to the First World War, are rooted in the reactions of the lifeworld to innovations and in retrospect seem quite hopeless and antiquated (Sieferle 1984).

It is not necessary, however, to ascribe to Marx a theory that this irresistible pressure for technological change is rooted in (his era's) modern technology, in the machinery itself. On the contrary, in his opinion it stems from the need of the capitalist system not only to extract and reinvest surplus value, but additional surplus value, in the trend towards extended reproduction. The fact that this dynamic can be released is primarily due to the fact that what we could call the sphere of labour, the economic subsystem, has become independently institutionalized and differentiated from all political and social relations which it was previously tied up in and limited by. Had a technology, or more precisely an institutionalization of technological change not been found, which was adequate for this kind of race to accumulate value, then the capitalist system would hardly have been able to assume a definite form. Seen in this way, capitalist society appears as a form of society that is capable of producing surroundings and a "second nature" that suits it to an extent which was previously inconceivable. The trend towards continuous technological change is therefore probably an absolute necessity for the existence of capitalism (a feature that Schumpeter has of course stressed as heavily as Marx, see Schumpeter 1961), but this trend itself has economic causes.

Needless to say, the capitalist relations of production existed before the onslaught of large-scale industry; this is an essential component in Marx's model of development. But it can hardly be doubted that it was in fact by means of precisely this kind of large-scale industry that the relationship between labour and capital became the fundamental class contradiction in modern society, that the modern or early modern societies became, up until today, inextricably connected with a capitalist class structure. From various vantage points it may also be of interest to point out that the world-historical significance of the emergence of large-scale industry and machine technology is that non-organic sources of energy are now becoming fundamental. Nevertheless there seem also to be grounds for maintaining that machine technology was the first technology that created a class — "created" in the sense of making the division into classes and the existence of an industrial proletariat a basic characteristic of society. Neither modern electronics, media technology nor modern pharmaceuticals have entailed any social revolution of *this* nature, even though the cultural effects have been revolutionary and they may be said to have led to a new phase within capitalism (Harvey 1989). It is therefore still appropriate to use the expression "the industrial revolution", even if it should be proved, as historians of economics are claiming today, that no qualitative leap occurred in the economic *growth* at the end of the eighteenth century.

Needless to say, Marx was not writing in a vacuum. (The following is based in particular on Maxime Berg's excellent account (Berg 1980).) The debate in England in the nineteenth century about the role of machinery had its starting point in Ricardo's theories about machinery. The debate had revolved around two points: Technological alienation and technological unemployment. In Marx's time the original alliance between the Tories and the preindustrial working class had broken down. Society was facing the unbounded advent of the factory system. Marx's general response to the misery that the factory system created was to distinguish between the specific capitalist use of machinery and the historical rationalization inherent in the development of the productive forces. Whether his own descriptions render possible another kind of use is an open question, to put it mildly.

Marx based his studies for the most part on the works of two theoreticians, Charles Babbage and Andrew Ure. Babbage (Babbage 1832) primarily interpreted the factory system as an extension of the preindustrial division of labour, while Ure ("The Pindar of the automatic factory", Ure 1861) basically saw the introduction of the machine from the angle of the profit motive and the capitalists' need to discipline and supervise the working class. Below I will deal with their influence on Marx's writings in greater detail.

In his description of the historical development of technology within capitalism in chapters 13 to 15 of *Das Kapital*, Marx operates with three forms of technological organization under capitalism: Cooperation, manufacture and finally the machine technology of large-scale industry. Formal subsumption as a period corresponds to the first form. Placing manufacture is more problematic. Cooperation, according to Marx, is the gathering of previously isolated producers under one roof, where they are organized by the will of one capitalist, even though they are technologically speaking working exactly as before. He splits manufacture into two forms (Marx 1965, p. 342): In the first, known as heterogeneous manufacture, the end product is a compound of the result of the labours of various specialized workers; the example Marx gives is the making of a clock. In the second, which he called organic manufacture, the end product itself goes through several different phases and stages of production; Marx gives the example of the production of needles, probably inspired by Adam Smith's famous example of the productivity-increasing function of the division of labour in the nail industry.

The form of increased productivity seen here is undoubtedly specifically capitalist. Indeed Marx sees it as a form of extraction of relative surplus value. Accordingly we should be confronted with a form of real subsumption (p. 364). The fact that Marx appears to perceive first and foremost the *machine* as the actual expression of real subsumption must be due to the fact that also the new forms of the organization of the production process within manufacture themselves take the form of a reorganization of a previous technology linked to the use of tools.

It is, then, in this context, according to Marx, "essential to hold on to" the following points:

> First, the decomposition of a process of production into its various successive steps coincides, here, strictly with the resolution of a handicraft into its successive manual operations. Whether complex or simple, each operation has to be done by hand, retains the character of handicraft, and is therefore dependent on the strength, skill, quickness, and sureness of the individual workman in handling his tools. The handicraft continues to be the basis (p. 338).

The organization of the labour operations in manufacturing grows out of the craft's own means of production. Its simple elements are the part-worker and his tools (p. 354). The logic of development of the tool, or perhaps more precisely *tool-using labour,* is therefore a vertical specialization, perhaps based on the need for new products, but certainly not horizontal in the combination of different types of tools to form new ones. The development of the means of production occurs within the framework of the tool by means of a double process:

> Manufacture is characterized by the differentiation of the instruments of labour — a differentiation whereby implements of a given sort acquire fixed shapes, adapted to each particular application, and by the specialization of those instruments, giving to each special implement its full play only in the hands of a specific detail labourer (p. 341).

Marx's statements that the change leading to industry is based on the development of the means of production and that the main task is to examine the difference between a machine and a tool (p. 371), are consequently to be interpreted to imply that the real subsumption is linked to this final revolution. In any case manufacture was after all a transitional phase: The role of the concept lies in a reconstruction of a stage of the development, and production according to the method(s) of manufacture was not a widespread phenomenon. Nonetheless we can perhaps see traces of an ambivalence here. If the revolution in large-scale industry is a result of *the means*, is it reasonable to say that the machine emerges from the manufacture, which is built on a division into specialized *labour* operations? This gains significance in connection with Winkelmann's criticism (see below). However, let us first take a look at Marx's description.

In general we could say that he views the logic of development of machinery as a break with the general vertical development, i.e. the increasing specialization, of the tool. Machinery presupposes a horizontal logic, a *combination* of elements. Once the basic forms of the tool have been given, as they have in historical time, technological inventions are strictly speaking related to the machine alone. Marx claims that regardless, the production of something new lies (only) in a combination of the elements that are already known. A study of the patents legislature in England and the USA appears to support this hypothesis. Up until 1850 a patent would be granted for an invention that consisted of a new application of previously known principles. After the middle of the century, however, the patents legislature became linked to the concept of genius. An American Supreme Court judgement from 1942 says of a new invention that it must "reveal a glimpse of creative genius" (quoted in Usher 1954, pp. 36–46).

According to Marx:

> All fully developed machinery consists of three essentially different parts, the motor mechanism, the transmitting mechanism, and finally the tool or working machine (Marx 1965, p. 373).

Marx regarded the machine tool as the starting point for the industrial revolution, the industrial events, which constituted the basis of his experience. It might however be most correct to say that he construed the development of machinery on the basis of an internal logic of development between the machine tool and the locomotion machine, a view that must be interpreted as Marx's summary of the industrial revolution, where he appears in particular to have found support for his ideas in the works of Charles Babbage.

The working machine or machine tool becomes the decisive element in machine technology — that part of machinery that constitutes the starting point for the industrial revolution. The role ascribed to the locomotion machine must be understood against the background of the machine tool. This is Marx's basic view, although he was not always consistent in his implementation of it. The keyword here is compensation, i.e. replacement of the human activity in the direct processing of the object. The machine tool is a collection of tools which perform the same function on the given material as the artisan did with his tools. The tool is not displaced by the machine.

On a closer examination of the working machine proper, we find in it, as a general rule — though often, no doubt, under very altered forms — the apparatus and tools used by the handicraftsman or manufacturing workman; with this difference, that instead of being human implements, they are the implements of a mechanism, or mechanical implements (p. 373).

Instead of being driven by energy from the worker's own body, as the tool is, the factory worker's machine tool is driven by means of an external source of energy.

Whether the motive power is derived from man, or from some other machine makes no difference in this respect. From the moment the tool proper is taken from man and fitted into a machine, a machine takes the place of a mere implement. The difference strikes one at once, even in those cases where man himself continues to be the prime mover (p. 373).

Once the tool is severed from the direct connection with the movements of the human body, the use of human power in the running of the machine tool itself becomes irrelevant. For the machine system to work to its full capacity, or 'reach its concept' if we want to be Hegelian, it is necessary that this happens.

Increase in the size of the machine, and in the number of its working tools, calls for a more massive mechanism to drive it; and this mechanism requires, in order to overcome its resistance, a mightier moving power than that of man, apart from the fact that man is a very imperfect instrument for producing uniform continued motion. But assuming that he is acting simply as a motor, that a machine has taken the place of his tool, it is evident that he can be replaced by natural forces (p. 376).

The machine is emancipated from the "organic barrier" (p. 374). So here we have the crux of the development of the real subsumption: The fact that the motion which drives the tool no longer directly, and after a while not even indirectly either, originates in the worker himself leads to the worker being reduced to an appendage to the machinery, as Marx constantly points out. Marx develops a pithy metaphor to describe exploitation through machine technology, and thus also through the factory system. As well as referring to the worker as an appendage, he uses metaphors from hydraulics: Machinery sucks or pumps surplus value out of the workers. Machinery can no longer be regarded as an extension of the human physiology, but rather it is this, the human physiology, that is forced to adapt to the machinery, as we have experienced in the endless waves of initiatives to rationalize labour, starting with Taylor at the beginning of this century.

Here we are not dealing with man's alienation from the essence of man, but from the very means of production. There is, however, also alienation on a new level in large-scale industry compared with the earlier capitalist modes of production. Since manufacture was based on crafts, there was no room for any application of science worth mentioning in production (p. 340). Science as an autonomous force of production (p. 361) "the universal product of the intellectual development" (Marx 1969b, p. 50), becomes along with machinery embodied in the constant capital. The working class finds itself confronted with the materialized expressions of significant parts of human intellectual history.

Marx thus regarded technologies first and foremost in their economic role, that is to say as means for the extraction of surplus value. It is important to bear this perspective in mind in relation to the criticism that has been directed at Marx's theory that machinery is essentially a compound of three different factors, and which to a great extent is based on Babbage. A. P. Usher attacked Marx on this point. Marx's definition, he claims:

> [...] ignores the distinction that can advisedly be made between mechanical elements and an extensive train of mechanism (Usher 1954, p. 116).

The main categories should therefore be regarded as the individual parts and those totalities that consist of them, where the individual part can perform totally different operations depending on the context. Usher's example is the screw, which can serve to join two elements, but which in a turntable serves to control the whole process (p. 117). Where Marx attempts to understand the development of machinery on the basis of the aforementioned logic, Usher claims that:

> The history of invention is more readily understood as a cumulative process if it is actually analyzed as a progressive development of larger and more elegant trains of mechanism, which are synthesized and composed by the arrangement of individual parts to achieve specific objectives (p. 117).

Usher finds support for his theory in Franz Reuleaux. His book, which was published as early as 1875, still exerts, it seems, a strong influence on all authors concerned with theoretical work within the structure and development of machinery. For Reuleaux a machine is:

> ...a combination of resistant bodies so arranged that by their means the mechanical forces of nature can be compelled to do work accompanied by certain determinate motions (Reuleaux 1875, quoted in Usher p. 66).

Economy and supervision of the power thus become the criteria for mechanical perfection. Reuleaux's most important idea was that the parts of the individual machines were no longer regarded from the point of view of their function within a whole, as they are when we speak of tool parts, etc., but that the machines were regarded as combinations of elements. The significance of this approach lies in the fact that it allows comparative analyses of different mechanical systems.

So how conclusive is this criticism? It is in any case not correct of Usher to claim that Marx was totally ignorant of this method of analysis. He discusses it in connection with the English mathematician Hutton, who, in keeping with Usher, regarded e.g. a screw as a machine, and concludes:

> From the economic standpoint this explanation is worth nothing, because the historical element is wanting (Marx 1965, p. 372).

Elsewhere Marx distinguishes between the different methods employed by technologists and mathematicians when considering the difference between a machine and a tool, and between labour machines and locomotion machines, and claims:

> For the pure mathematicians these questions are of no concern, but they become very important, in connection with the demonstration of the relationship between the conditions of human societies and the development of the material mode of production (Marx 1863, p 321).

It is therefore a question of whether we are not here seeing the outline of an interesting typology for analyses of technology on different levels of attachment to economic affairs. On one level there is the pure description of technology as a system. Hutton's and Reuleaux's relationship to technology is without a doubt coloured by their consistently exclusive and engineering approach to technology. On another level it becomes possible to make diachronic as well as synchronic studies of the relationship between these mechanisms and different stages of development in the predominant modes of production. It is on this level of Marx's theories that this account has moved so far. A third level would be an explanation of the way in which the technological choices of the capitalists, i.e. choices between substitution possibilities within the fixed capital, were influenced by economic and political circumstances. It is probably on the latter two levels, where we find room for "a critical study of technology", that Marx wanted to contribute, and it is to these levels he ascribes this kind of meaning (p. 372, note 3).

This is quite clearly an invitation to do research on Marx's part, which has generally been ignored. According to many commentators, the Norwegian historian, Kristine Bruland, is one of the few who have taken up the gauntlet. She describes the field thus:

The general point here, as I see it, is not simply that innovation may occur in some cases to destroy production bottlenecks created by the recalcitrance of workers. It is rather that problems of technical innovation — and problems of the development, application and structure of technologies more generally — in capitalist economies, should be located within a clear conception of the social and managerial field of forces underlying capitalist production, and which structure many decisions relating to technique. The calculations of entrepreneurs should be conceived as including this kind of component (Bruland 1982, p. 117).

A more weighty criticism of Marx's representation than that put forward by Usher was penned by Rainer Winkelmann (Winkelmann 1982) — more weighty because it actually has its starting point in the relationship between economy and technology. In short, Winkelmann tries to illustrate that Marx had placed too much emphasis on Babbage's interpretation of machinery as a synthesis on the basis of the specialization, i.e. on the division of labour in manufacture. This is to overemphasize the concept of the division of labour in Smith's classical economics, which the technologist Babbage strangely enough has followed here. But this principle does not apply to the machinery of the English textiles industry, and it is the development in this field, particularly because it paved the way for potential uses of steam power, that is of decisive importance. The labour that was replaced by machines was in fact *already at the outset* complex. There is thus no empirical evidence to support the existence of the process assumed by Babbage, and following him, Marx. Berg too underlines the fact that Babbage's view was already quite antiquated at the time at which Marx was writing.

Elsewhere Marx follows Ure's approach more closely. His actual analysis of the function of the machine, based on the relationship between its three components clearly points in the direction of Ure's perspective that the essence of factory labour lies in the fact that the labourer is subordinate to a central source of energy, that the energy employed does not originate in the body of the worker. Perhaps it was Marx's disdain for Ure's unctuous apologia for capitalism that caused him not to acknowledge this inheritance more directly. Regardless of Winkelmann's justified criticism, however, which in fact calls into question the sketch of development rather than the actual definition of the machine, Marx's theories about machinery and the factory appear to be model examples of what we could call an analysis of the materiality of society.

I am now going to illustrate Marx's point more concretely through a discussion of the theme: What was the essence of the factory system, the technical innovations, that rendered possible the real subsumption, or one more social moment, labour discipline? I state once more for the sake of clarity that this is not intended as an argument to support some form of technological determination. *Both* the introduction of labour discipline *and* machine technology are only comprehensible in the light of the double nature of the production process, i.e. as a process of labour and as a means of exploitation. My only aim here is to illustrate the significance of the concept of materiality. I am intervening here in a debate the theme of which seems to be: What was the most important feature of the new modes of production, the technology or the social organization? Kristine Bruland (Bruland 1989) defends placing the emphasis on the social and disciplinary factors behind the emergence of an industrial working class. Based on what I claimed above, namely that the importance and fruitfulness of Marx's theory lie in the fact that technology must be viewed in the light of the aspect of exploitation, it should be quite clear that I have no prejudices against the social aspect. I do however have my suspicions about the validity of the approach machinery versus discipline. The aim of the following is to show that there is no element of the chicken and the egg in claiming that it was machinery that paved the way for the practical implementation of factory discipline, the new organization. The fact that the capitalists' attempts to enforce discipline commenced before industrialization does not therefore constitute a counter-argument; the question with regard to my approach is *when* discipline could be implemented.

Marx claims in *Das Kapital*:

> Since mechanism of manufacture as a whole possesses no framework, apart from the labourers themselves, capital is constantly compelled to wrestle with the insubordination of the workmen (Marx 1965, p. 367).

Marx is then saying that it was extremely difficult to implement any really modern form of labour discipline in manufacture, since the necessary basis in the actual process of production was absent. The implementation of discipline was dependent on subjective factors, also on the managerial level.

Neil McKendrick provides a good example of this kind of struggle in an article about Josiah Wedgwood's pottery company (McKendrick 1966). According to Marx's terminology, Wedgwood's famous plant for the production of pottery was a manufacture rather than a factory, and it was the scene of a battle of this very nature between a prominent representative of the new entrepreneurial class and skilled workers. The workers had to be purged of previously engrained habits, and wrenched out of a labour routine they had adopted as independent workers and craftsmen. Wedgwood demanded of them:

> [...] punctuality, constant attendance, fixed hours, scrupulous standards of care and cleanliness, avoidance of waste, the ban on drinking. They

did not surrender easily. The stoppages for a wake or a fair or a three-day drinking spree were an accepted part of the potter's life — and they proved the most difficult to uproot (McKendrick 1966, p. 70).

For David Landes these kinds of deliberations seem to be decisive for the determination of the essence of factory production. He rejects two other possible criteria: The distinction between work and capital as a criterion of factory production, because this existed much earlier; and the application of machinery, because this does not automatically lead to production on the premises of industrial principles. Marx would have agreed with the first, and he too notes that machinery may be used within manufacturing. But Landes argues further:

> Others have seen the essence of the factory in the use of machinery and the introduction of a central source of power, whether water wheel or steam engine. This interpretation comes closer to the facts [...] Yet the use of a central power source does not necessarily entail recourse to factory organization. [...] No, the essence of the factory is discipline — in the opportunity it affords for the direction and coordination of labour. [...] It took several generations to form a punctual, efficient factory labour force. And nothing conveys better the economic contribution of the new arrangements than Neil McKendrick's essay on the Wedgwood factory — an example the more impressive precisely because this was as yet an unmechanized enterprise (Landes 1966, p. 14).

The work discipline which Wedgwood enforced does of course bear some resemblance to industrial discipline, but it was not identical to it. In Wedgwood's case the main goal was to get the workers to turn up at regular times every day.

> Wesley and the Sunday schools taught the industrial virtues [...] Wedgwood's factory discipline — his bell, his embryonic clocking-in system, his rules and regulations — insisted on them (McKendrick 1966, p. 89).

This is also necessary in large-scale industry, but it does not suffice. In large-scale industrial production, labour discipline is incorporated into the actual labour process itself, as this in turn is dictated by the machinery. The enforcement of discipline in manufacturing is and must be dependent on the personal charisma of the supervisor, the capitalist, which can only be a means of strengthening this internalization. It is in fact difficult to imagine that this *could* be exercised directly under the conditions of production that prevailed in factories in the nineteenth century. Nobody could do the rounds then writing "This is not good enough for Josiah Wedgwood" on the products.

Manufacturing requires discipline based on the organization of labour power, but offers no clues as to how to go about implementing it — indeed its nature makes it practically impossible. The factory system with its mechanical production both requires discipline and generates it in the very work process. Only production with machinery provides the material foundation for the permeation of labour power with capitalist rationality, or more precisely, that form of rationality that capitalism requires of the workers. I see this as evidence for my thesis: It is dubious to call discipline the essence of the factory system because it is actually the machinery that *provides* discipline with a foothold, allowing this rationality to be enforced upon the labour power.

I am therefore against drawing a parallel between Wedgwood on the one hand and the establishments of Arkwright and Boulton on the other, as McKendrick appears to have done. Marx builds on Ure, who consistently praised Arkwright. Arkwright's plant from the 1780's was a factory based on water power, not the workers' own energy. In Ure's view this constitutes the very definition of a factory: A factory worker is "a person who in the labour process is subordinate to the moving power in the production process" (Ure 1861, p. 8).

It is worth pointing out here that Landes too, when describing the new factory system in detail, had to fall back on the conditions for the actual performance of the labour in order to define the difference from the previous, highly organized units of production, e.g. iron smelting in the eighteenth century.

> Should such units be designated as factories? From the standpoint of the two critical criteria — concentration of production and maintenance of discipline — the term certainly fits. At the same time, they differed in one important regard from the textile mills that were in fact the prototype of the factory as we know it: however thoroughly the work in these forges and yards was supervised, the pace was set by men and not machines. It was spasmodic rather than regular. There were moments that required a burst of concentration and effort [...] And there were quiet moments, while the mix boiled or the men waited for the next piece to be ready (Landes 1970, p. 121).

It is the logistical problems attached to the application of labour power which are central here: Attempts to shorten unnecessary breaks are a major concern of the supervisor. The very actualization of the subjective conditions of the labour remain unchanged.

Mantoux's classic account expresses the same interpretation. Admittedly he rejects the parallel between manufacturing and industry, on the grounds that the economic significance of the former system was negligible (Mantoux 1961, p. 30). His underlining of the importance of the necessary connection between the factory system and machinery, however, is without any ambiguity.

> The use of motive power other than the muscular strength of men or of animals is one of the essential features of the modern factory system. Without it, though there might have been machines, yet there could have been no machine industry [...] The gap between manufacture and the factory could in fact never have been bridged (p. 311 f.).

It is on this basis that the organization of labour power in the period must be explained.[13]

> It [the old water-wheel T.K.] enabled work to be organized in large workshops, where the men were brought under that strict discipline which was the necessary and immediate outcome of the machine industry (p. 312).

Towards the middle of the nineteenth century the use of mechanical power was generally considered the decisive characteristic of factories. Ure has already been mentioned above, but this use of language also caught on in legal circles:

> The word factory...shall be taken to mean all buildings and premises...wherein or within the close or curtilage of which steam or any other mechanical power shall be used to move or work any machinery [...] (An act to amend the Laws relating to Labour in Factories. June 6 th. 1844 quoted from Mantoux p. 39).

This is not meant to settle this debate once and for all. This outline is primarily intended as a demonstration of an application of the concept of materiality.

---

[13] Cf. also p. 26, and in particular p. 37 ff. where he supports Marx's use of machinery to distinguish between manufacturing and large-scale industry. It is significant that Mantoux generally discusses discipline in connection with the real industry of cotton production.

# 5 Jean-Paul Sartre

Sartre's theories on technology, as developed in the tome *Critique de la raison dialectique (Critique of Dialectical Reason)* from 1960 (Sartre 1976, henceforth abbreviated to *CRD*),[14] play a special role in this book. I have borrowed from Sartre (not least via Dag Østerberg's development of Sartre's ideas) the concept of technology as social materiality, as the practical-inert field which intercepts and influences all practices in society. This form of inertia or alienation of man's activities and their results are, in Sartre's view, an aspects of more than just technology in the traditional Marxist sense of the means of production. Such inertia is also an aspect of technological elements in what we could call the lifeworld, e.g. media technology and public transport — in other words, it is an aspect of materiality in the very widest sense of the term. Sartre's analyses of the radio listener and bus queues are examples of nuggets from professional philosophy (if it is possible to call Sartre a professional philosopher) that have become embedded in the common cultural consciousness. So in this respect at least, Sartre's theories can be said to go further in a productive way than Marx's.

Sartre uses this idea of the inertia and alienation that will descend over all human activity, or which has actually already done so in all history known to us, not only to describe man's relationship to technology and *to* production and *in* production, but also to describe the development of intersubjective relations and different types of collectives. The discussion of technology in a more traditional sense is part of a wider theory in which all configurations between people, and not merely those which by means of labour are directly aimed at the processing of previously processed matter, have this inert character. They are reduced to something which is no longer subjective, they are conquered by alienation and are forced to get hopelessly engaged in a battle against this alienation, or seriality. This is Sartre's term for this phenomenon when it appears on the level of intersubjectivity. Sartre's description of these various types of alienated collectives, as a series, a group, etc., is not however my main theme here. I will only be mentioning these themes if and when they can help to shed some light on his treatment of technology, the practical-inert field proper.

---

[14] The quotations below from this book refer to the English edition Sartre, J.-P. *Critique of Dialectical Reason* (Sartre 1976).

In order to elucidate Sartre's method for approaching technology, it might be advantageous, by way of an introduction, to look at the relationship between *CRD* and some of his other works. I am thinking here first and foremost of the relationship between this relatively late work and the classic within existential philosophy, *L'Etre et le Néant (Being and Nothingness)* from 1943. We appear to be confronted with an insurmountable incompatibility between the unshakeable doctrine of freedom in the first work and the emphasis on the way that matter binds human intentionality in the second. Marx has been added as an important precondition and despite the Hegelian jargon that permeates *CRD*, it is apparently not Hegel as a theoretician of consciousness, but rather Hegel as the philosopher of alienation, of the state, of the institutions and of labour, as the author of the *Jena Manuscripts* and of the *Philosophy of Law*, whose constant presence can be felt in the background.

The majority of Sartre biographers claim that he did not turn his attention to politics until after the Second World War, and that his involvement in Marxist Politics only really developed around the time of the Korean War. Nevertheless Sartre seems always to have been fascinated by the relationship between his own original ontology of consciousness and a Marxist theory of society, a theory which he wanted both to develop further and at the same time liberate from its naturalistic and other perversions of a dialectical nature.

As early as in 1938 he wrote that such a fruitful hypothesis of labour as the materialistic conception of history did not need "an absurdity such as metaphysical materialism", i.e. the leading dialectical materialism in the communist parties. In 1947 he described history as a combination of objective and subjective elements, where the dialectic of human actions is met by a counter-dialectic, a meeting which all the same lends history a certain dialectical feel. (For these quotations, see Jameson 1971, p. 206). This is, of course, the very position and function that inert matter, the practical-inert field, would come to fill in *CRD*. What we encounter here is the idea that materiality constitutes a counter-practice to the projects of man.

In commentary literature we find what appear to be several different views of the relationship between Sartre's original position in 1943 and *CRD*. One interpretation accentuates the criticism Merleau-Ponty directed at Sartre in Merleau-Ponty 1968, chap. 5: That he, despite his wish to give a phenomenological account of the position of the human consciousness in the world, came to a standstill at a kind of Cartesian dualism between a Being-in-itself, understood as total immanence, as being completely shut within oneself, and the consciousness, a being-for-itself, a form of pure negativity. *CRD* was then intended as an attempt to search beyond an analysis that moved only on the level of theories of consciousness, in order to emerge itself in politics and society. (For this kind of interpretation using Merleau-Ponty as a starting point see in particular Caws 1977 and Aronson 1980. The accusations against Sartre of voluntarism and a deficient sense of the objectivity of political reality were seen in the light of this desperate wish to enter the real world once and for all.

Another interpretation perceives *CRD* as a more immanently philosophical project, the aim of which was to master the internal deficiencies in the theory from *L'Etre et le Néant*. According to this interpretation Sartre's conception of the body as the situated consciousness, is irreconcilable with the conception of the consciousness as pure negativity. Only *CRD* paves the way for the concept of passivity and affection that this kind of theory of a bodily subject must also encompass (Dreyfuss & Hoffmann 1981, p. 235). It is therefore reasonable to add that this theory of the bodily subject is in turn introduced into Sartre's thinking through the theory of material practice: This bodily subject is first and foremost a *working* subject.

There is of course absolutely no reason to view *CRD* exclusively as an internal or external revision of the original position. Nowadays it is perhaps equally fruitful to regard *L'Etre et le Néant* in the light of the later work, as Fredric Jameson and possibly also Dag Østerberg suggest (Østerberg 1993). The concept of freedom in situations in the first work can thus be understood as an as-yet abstract version of the description of practice in relation to the surroundings which appears in *CRD*, at the same time as those forms of interhuman struggle and relations of power, especially those attached to the analyses of the look, which Sartre so energetically worked out in *L'Etre et le Néant*, now acquire a more secondary position in relation to the position of the individual in production and the class struggle.

I find evidence which supports an interpretation in this direction in one of the few passages where Sartre himself mentions in more detail the internal philosophical relationship between his two main works. This passage makes it clear, as I mentioned earlier, that the systematic place for a passivity which Dreyfuss and Hoffman found, was only introduced because Sartre perceives the bodily subject as working. This is what he alludes to by the term practical action. If being-for-itself appears as being-in-itself this is due to the fact that:

> ...the very structure of action as the organization of the unorganized primarily relates the For-itself to its alienated being as Being in itself. [...] Certainly, *praxis* is self-explanatory *(se donne ses lumières)*; It is always conscious of itself. But this non-thetic consciousness counts for nothing against the practical affirmation that *I* am what I have done [...] Fundamental alienation does not derive, as *Being and Nothingness* might mis-

lead one into supposing, from some prenatal choice: it derives from the unequivocal relation of interiority which unites man as a practical organism with his environment (Sartre 1976, p. 227).

Still, it may obviously be claimed, as Jameson also points out, that Sartre in *CRD* is still describing interpersonal relations such as class identity and class struggle according to his model of interpersonal analyses in *L'Etre et le Néant* and his analysis of Jean Genet's childhood. The classes constitute themselves according to a model of intersubjectivity through and for each other (Jameson 1971, p. 301). The clearest example of this is Sartre's contention that the triad, as opposed to the dyad, is the fundamental form of social connection: He describes how a third person, a tourist, sees the two originally separate and different workers. Admittedly an essential change occurs, as Klaus Hartmann points out, when we reach the more complex forms of social organization in *CRD*. The medium for this constitution of the classes is no longer the look, as in Sartre's first classic, but rather they are primarily constituted in and by means of materiality (Hartmann 1966, p. 79). In order to illustrate just how interwoven the theory of classic materiality and the old theory of consciousness are, however, it might be useful to indicate another factor. Sartre's claim that the triad is fundamental, that two individuals cannot constitute an entity, is based on the fact that on meeting each other the two subjects will attempt to turn either himself or the other into an object. A meeting of two on the same level can only occur when an outside third party constitutes them for himself as two subjects (cf. Knecht 1975).

Now, it is not of vital importance in the context of my book whether or not the doctrine of social materiality can be made to agree with Sartre's ontology and the doctrine of consciousness of his early phase. The point is that despite all the displacements in perspective, it is barely possible not to recognize certain *premises* from the original ontology in *CRD*. Whereas *L'Etre et le Néant* placed the bulk of the emphasis on the concept of negation and man's attempt to conquer the moment of nothingness in himself and his experience of the world, *CRD* has its starting point in the need, and then moves on to scarcity as the basis for the existence of man (Jameson 1971, p. 232). Admittedly in this work Sartre is no longer dealing with the idea that we, on the basis of the basic structure of our consciousness, actually want to turn ourselves into things, but rather that the things, i.e. the processed matter, turn us into things, owing to features of our practice that must be apprehended as a basic characteristic of our history up until now. In the same way as the secret of existence in *L'Etre et le Néant* was the tendency of our consciousness to turn itself into a thing, it now becomes the tendency of our practice to transform itself and us into inert entities.

More over, on what may be an even more basic level, we recognize an intuition in Sartre that might almost be called Manichaean: Freedom is and remains escaping being, which here takes the form of the inert material and the seriality of the group.

In order to attain a better grasp of the structure of *CRD* it might be useful to look at its relationship to *Question de Méthode (The Problem of Method,* Sartre 1963) originally from 1957, but which was published as the foreword to the French edition of the main work from 1960. We are primarily interested here in the extent to which *CRD* uses the progressive-regressive method which Sartre developed in this earlier work. It seems to me that neither exclusively nor first and foremost is this method intended to place the individual in a historical context. I would rather place the emphasis on the fact that the method aims to illustrate what is unique, the irreducibly individual elements of any human project. We could even say that we are being confronted with a *lived* reality here, the reconstruction of a reality as it was originally seen through the eyes of a single individual. The associations with the German historical school and Dilthey, with early hermeneutics, are unavoidable. It is nevertheless important to remember that Sartre was not interested in reexperiencing, but rather reconstructing a historical process as it appears to an actor, as the life of this actor.

I cannot deny that there are passages in *CRD* that may appear to be examples of this kind of method. Nor is it unreasonable to say that the scarcity and the inertia of matter are factors to which we return in order to understand the significance of an action. Nevertheless it seems quite clear that this is not the method of *CRD*, despite the fact that *Sartre 1963* was printed as a foreword to it. Instead it is the method we find in Sartre's biographical studies, such as the studies of Flaubert, that is used in *CRD*. Indeed in *CRD* it is Flaubert he uses as an *example*. An interpretation of *The Problem of Method* along these lines seems, as we shall see, also to be the most fitting for the entire structure of *CRD*.

Sartre distinguishes between 1. The preconditions for a structural and historical anthropology, 2. A regressive procedure which underlies sociology, and 3. A progressive procedure which renders historical knowledge comprehensible. It is also said that Volume 1 of *CRD* stops before history, that is just the first moment (Sartre 1975, p. 71). (What Volume 2 deals with is not clear here, but it is hardly of vital importance.) It seems reasonable, then, to join Mary Warnock in stating that the objective of *CRD* is in fact to *give the grounds for* a regressive-progressive method, not actually to apply it (Warnock 1982).

I therefore take *CRD* to be a transcendental philosophical investigation, in the precise sense of the term. Its aim is to map out conditions of possibility. Furthermore it is a transcendental investigation which unites the two main concerns in Kant's critique of reason, namely to describe the conditions of possibility of a field of objects and the conditions of possibility for understanding it, or to use Sartre's vocabulary, for intelligibility. The field of objects with which we are confronted here is history, and that particular form of intelligibility, or form of knowledge which is necessitated and made possible here, is called dialectics. The counterpart to history is nature, and the counterpart to dialectics is called analytical thought. (This should be taken to mean naturalistic or ahistorical forms of thought such as the dialectic of nature, since Sartre ignored analytical *philosophy*, although there are some likenesses, as we shall see.)

This is however also a Hegelian form of transcendental philosophy: The basic conditions of possibility are not given ahistorically, like Kant's categories, but unfold in history. The three main elements in the structure of the book are: 1. The dialectic that constitutes the individual practice, 2. The principle of practical inertia in groups and in inert matter, which constitutes an anti-dialectic to this, a passivity which opposes the original practice, and 3. The group that constitutes a new dialectic (Sartre 1976, p. 66).

Of course nothing is more inevitable in Hegelian contexts than having to count to three: The reader will therefore be relieved to know that there is no division in Hegel's works that corresponds exactly to Sartre's, but they do share one basic intuition: Alienation can only be overcome through a form of more or less consolidated intersubjectivity. In this way an element of criticism in the Kantian sense of the word is supplemented in *CRD* with a moment of criticism in the sense it acquired in a later dialectical tradition: *CRD* is a critique of the validity of forms of thought in Kant's original sense of criticism and of the legitimacy of forms of reality in the sense of criticism we find in Hegel and Marx.[15]

Sartre's constant use of the expressions 'totality' and in particular 'totalization' in connection with man's creation of his history and the possibility for comprehension of this, leads us inevitably to the question of whether and to what extent Sartre has been influenced by Hegel's notion of a final and absolute understanding of history. Furthermore, despite the fact that Sartre in true, left-Hegelian style remains attached to the idea of the unity of theory and practice, he seems to have conceived of the possibility of a final and global control over and insight into the entire history of mankind. He does of course use the expression totalization in so many rather restricted contexts that connection with Hegel's idea of an absolute spirit is at any rate not the only meaning of the expression in *CRD*. It seems more reasonable to

---

[15] Hartmann's work is a commentary which places the main emphasis on Sartre's relationship to Hegel, in particular Hegel's logic, (Hartmann 1966).

join Ingbert Knecht in regarding the concept of totalization against the background of Sartre's phenomenological assumptions, which for him presuppose an individualistic or, if you prefer, a nominalistic point of departure (Knecht 1975). Totalization will then only refer to the way in which the individual's projects become joined together and produce a result on the collective level, based on the forms of processed matter within which they develop, as in the example of the Chinese farmers below.

As regards this point there are grounds for looking at another author for further illumination of Sartre's way of thinking, a writer who has been overlooked by all the philosophical commentators with whom I am familiar, namely the French historian Fernand Braudel. Sartre does of course refer to in his book *Le Méditerranée et le monde Méditeranéen à l'epoque de Philippe II (The Mediterranean and the Mediterranean World in the Age of Philip II*, quoted here from the English edition as Braudel 1972) in connection with his own analysis of the Spanish import of precious metals in the sixteenth and seventeenth centuries, to which I shall return later. There is evidence however to support the claim that Braudel's methods made a greater impression on Sartre than this single reference would imply. Sartre's own concept of totalization, as the unity that arises through and links the projects of various individuals, is not only rooted in Hegel's category of totality, but also in Braudel's discussion of the Mediterranean areas. Braudel interprets it at first as an entity assembled from various factors, united in the first instance by geographical considerations, "a sea surrounded by mountains" (Braudel 1972, p. 25), and later as a network of demographic, commercial, political and institutional factors. The Spain of Phillip II is indeed itself described as a totalized entity "a container" (which leaks, like a sieve) for the American gold and silver, a container which consisted of natural and institutional barriers: Geographical limitations, customs barriers and mercantile politics (Braudel p. 476). Here we catch a glimpse of the concept of materiality of which Sartre gives a phenomenological interpretation, a matter which consists of both the processed natural landscape, as it appears with its given preconditions to new techniques "today Florence is by the sea...", and the manufactured artefacts, tools, machines and technological systems (p. 178).

We now know the position of technology or social materiality in Sartre's view of history; it constitutes a counterpart *within* the historical context to human practice and to genuine, i.e. not alienated, intersubjectivity and formation of collectives. The principles for this counterpart, this field and its principles, are dialectical because they, in spite of all alienation, are situated within a man-made history, and they are anti-dialectical because they cross, distort and make unrecognizable, i.e. make inert, the human intentionality which is embedded in the processed materiality.

As we have already seen, Sartre begins his account of man's practical and active relations to his surroundings by describing the organism's urge to satisfy its needs. This does not necessarily have to be taken as a form of biologism or a reduction of social relations to merely natural conditions. It is more reasonable to say that Sartre wanted to develop the conceptually basic structure of the labour of the individual on the basis of and by means of modifications of this relationship to nature, which already contains the seeds of more complex relations. We can find a parallel to this approach in Hegel's exposition of life in Hegel 1953, where he presents the structure of the self-consciousness and the spirit for the first time. The organism is confronted with nature, its surroundings, but the relationship only becomes dialectical when labour, which then negates this distinction, enters the scene. On the one hand labour is a project that constitutes the world and all its components in the light of this type of project, while on the other hand we see here the original entity of activity and alienation.

> Its [human labour's] primary movement and its essential character are defined by a twofold contradictory transformation: the unity of the project endows the practical field with a quasi-synthetic unity, and the crucial moment of labour is that in which the organism makes itself inert (the man applies his weight to the lever, etc.) in order to transform the surrounding inertia (Sartre 1976, p. 90).

We could even say that for Sartre labour is the basic pattern of the dialectic, and is thus also a model for all reflection, for those forms of intelligibility he will come across later.

> This inert materiality of man as the foundation of all knowledge of himself by himself is, therefore, an alienation of knowledge as well as a knowledge of alienation (p. 227).

Sartre also attempts in *CRD* to understand both the formation of collectives and intersubjectivity, and even the very nature of language, on the basis of this model. This lends his philosophy of language an entirely unintended flavour of instrumentalism, which makes it inadequate as an explanation of the function of language as an intersubjective medium. We are forced to conclude that Sartre's point of departure renders impossible a satisfactory theory of intersubjectivity. Habermas accuses Marx of absolutizing a purposive-rational perspective on historical change in Habermas (Habermas 1987, chapters 2 and 3). We can of course object that even though Marx was also working within the framework of subject–object philosophy and perceived the genesis of the individual according to a model of reflection, he was nevertheless primarily interested in analysing the various and specific historical configurations of technical and social categories. Sartre is therefore hit even harder by Habermas's criticism, especially if we continue to perceive this theory just as it was intended, i.e. as a theory of the basic preconditions for the comprehensibility of history. He does however have a striking approach to one sub-aspect: The material mediation of alienated mediations. But when reflection is understood through the basic model of labour, all preconditions for a concept of a differentiated rationality that enables us to criticize this alienation disappear.

The individual practice in the processing of nature thus assumes the position of the point of departure for all historical phenomena in Sartre's view. It does not however determine the decisive character of any specific period. It does not even constitute any restrictive condition for the course of history as a whole. This kind of anthropological basic condition, which *CRD* set out to reveal, is only achieved when *scarcity* is introduced.

This category is probably one of the most controversial and at first glance astounding elements in *CRD*. It is undeniably tempting to perceive it as a moment of a Hobbesian or perhaps even a neo-Malthusian inclination in Sartre himself. Moreover it is scarcity that introduces evil and Manichaeism into the world. But on the basis of the transcendental philosophical interpretation of the work, which I have outlined above, I prefer, at least at this early stage, to reject interpretations in this direction. Scarcity is a structure of understanding or intelligibility in all history as we know it. It is of course a real phenomenon which expresses itself as hunger and want, but it primarily appears as the point of view that must be used as a starting point in any attempt to understand the course that the history of mankind has actually taken. The historical phenomena we study only become comprehensible if we perceive them as structured by the omnipresent scarcity. Seen from this angle, scarcity finds its place among those structures which Sartre developed in order to be able to fulfil his main objective: The development of a transcendental anthropology.

Admittedly scarcity as a category is less universal than labour, which for Sartre is the precondition for all possible history and all historical understanding. Since scarcity is therefore not the precondition for history as a whole, but merely for the specific features which characterize our history and the type of people we know, it strictly speaking hardly qualifies as an a priori structure. This does not however constitute a conclusive objection to calling Sartre's project transcendental: The analytical transcendental philosophy of this century, as represented by Peter Strawson, also confines itself to finding the necessary conditions of possibility for our actual experience, as opposed to all possible experience, as Kant originally strove to discover. Nor is Sartre's approach deductive: The concrete is not already contained in the abstract arrangements. We cannot understand the actual development of history on the basis of scarcity alone. Furthermore our starting point is not the conditions of possibility, but rather actual history, in order then to search back for its conditions of possibility. This exemplifies the *practical application* of a progressive-regressive form of representation within transcendental anthropology itself. Once again we notice a striking resemblance to analytical transcendental philosophy. Strawson's form of transcendental philosophy has as its goal to find the necessary preconditions for the distinctions with which we actually operate in our language, not in all languages in general.

Scarcity alone does not, then, constitute the contrafinality of the inert field, for this phenomenon was, as we saw above, already linked to labour (p. 123). What it does constitute is Man in our history, or, to be more precise, the distinctive form that all interhuman relationships acquire in a social field which is marked by scarcity, by alienation. We have all become an other to each other and ourselves. The impact of scarcity on labour, which for Sartre remains the basic category of history, is a specific modification, that is, it is performed by "the people of scarcity".

> But given a social field which is defined by scarcity, that is, given the historical human field, labour for man has to be defined as *praxis* aimed at satisfying need *in the context of scarcity* by a particular negation of it (p. 136 §7).

He later says of this very negation that it forms the grounds for the comprehension of history and assumes the shape of a practical dimension of non-humanity (p. 146). This inhumanity is of course that form of alienation towards each other upon which we have already touched.

On this level it would appear that Sartre at any rate *exemplifies* scarcity with real physiological states of want, such as undernourishment and absolute impoverishment. Perhaps the problem is not primarily, as several commentators have claimed, often with reference to Marshall Sahlins's *Stone Age Economics*, that early societies such as those of the Neolithic period had a surplus of goods rather than a distinctive character of scarcity. A better example is perhaps that at the same time as large parts of the world are in fact exposed to scarcity in the physiological sense today, absolute impoverishment has ceased in the core of capitalist countries. Neither is a strictly Malthusian interpretation necessarily the only possibility on this point. Sartre's basic formula: "There is not enough for everyone" can be given a broader meaning. Scarcity implies that everyone's plans are crossed by someone else's. For example, two companies cannot both gain a complete monopoly of the market, nor can two multinational concerns attain the full right of disposal over a country's sources of raw materials. A scarcity of resources such as time and information when decisions have to be made is actually just as a good an illustration of Sartre's point.

It is incorrect to say that because labour is modified by scarcity, the projects of the individual producers are totalized in a non-intended form, and with a non-intended result. The phenomenon of contrafinality was, as we saw, more fundamental than that. The modification of labour by scarcity thus entails a new category, the *inert* nature of the processed field. It is this inertia inherent in processed matter that preserves the reciprocal alienation or otherness between "the people of scarcity", and which turns the productive forces into creators of and impact areas for distorted human relations.

In other words: we have seen how production establishes itself and how it makes alterity a characteristic of the relations of production in the context of scarcity, or the negation of man by materiality as an inert absence of matter, and we shall now investigate how alienation becomes the rule of objectification in a historical society to the extent that materiality, as the positive presence of worked matter (of the tool), conditions human relations (p. 152).

Sartre observes that the potential for class contradiction grows out of this relationship. I am not going to continue to explore this train of thought here. In connection with a theory of technology it is more important to delve more deeply into the other elements which make up the concept of inertia.

We can isolate one fundamental characteristic straight away, a feature which we have already dealt with above: The inertia of materiality is first and foremost associated with a form of anti-practice or passive synthesis. Through this the processed matter comes into conflict with, and exercises its own activity on and over the practice of man. Sartre formulates this point of view in paradoxically anthropomorphic terms, such as when he appreciatively cites Mumford on what the factory and steam power *require* (p. 159). This is only meaningful if we bear in mind that Sartre constantly presupposes that we are confronted with processed matter here, i.e. that what we come across as the anti-practice of matter is the embedded form of other people's practice.

Owing to the chaotic nature of the text, it is easiest to latch on to the occasional concrete examples which Sartre himself uses. The first and simplest example is that of the deforestation of the Chinese culture landscape (p. 161 f.). Each individual farmer felled a number of trees in order to obtain more cultivable land. This of course entailed not only an increase in the area of agricultural land, but also a change in the geographical conditions in such a way that China suffered constant floods. What we encounter here is above all the phenomenon of contrafinality: The result of one individual's project does not only render impossible or reduce the possibility for the fulfilment of the original intention. The result is not only different to what was intended, but is in fact harmful to the actors. But once again this contrafinality only arises as a result of and as a cumulative effect; it is a result of everyone in isolation following the same strategy of deforestation. It is also worth noting that the floods in China only occur because of certain prevailing hydrographic and geological conditions, as Sartre himself was quick to point out. He is undoubtedly right in stating that these conditions *only* became a threat, a potential catastrophe for the farmers, through their deforestation. These conditions thus only become relevant *as data* within the framework of the explanations provided by historical knowledge. It does however seem that this knowledge remains scientific, i.e. analytical in the vocabulary of Sartre. This would appear to be at least one good reason for doubting the contention that history as a discipline presupposes and only uses a specifically dialectical form of knowledge.

The example of the inflation in Europe which was created by the import of precious metals from America is somewhat more complex. In the first instance it is connected to the fact that increased wealth apparently leads to economic catastrophe through inflation and thus a rise in nominal prices. The gold, and especially the silver imported from America, were melted down and made into coins. Spain was not only a container, but a melting pot. In their natural state these metals represented wealth because gold was considered valuable in itself, and in their minted form they still represented wealth, since they consisted of precious metals, and they were also regarded as a sign of wealth — they represented an amount of value. The metal value of a gold coin and the amount of wealth it represented as a coin did not of course have to be equal. By means of this process the amount of money naturally increased because more and more gold was minted, creating a galloping inflation. According to the premises of mercantilism, the money was in itself wealth in the eyes of the individual merchant or banker. It was rational to acquire as much money as possible, but as the demand for minted precious metals grew, the value of each individual coin and each accumulated sum of money fell. This led to a capitalist strategy with a decrease in wages in an era with rising prices and general social misery.

The social materiality with which we are confronted here is in fact the precious metals themselves, and the physical processes and institutional limits which surround their conversion to coins, i.e. the Spain which was a melting pot for precious metals. This is not a simple case of contrafinality, but rather a phenomenon reminiscent of Marx's concept of fetishism and his criticism of the idea that circulation alone can lead to a general increase in wealth. The mercantilist actors were not capable of distinguishing 1) between gold as a natural phenomenon and gold as valuable in a human society, and 2) between gold as a valuable phenomenon and minted gold, which itself is in turn both a) wealth, i.e. a bearer of a purely social characteristic, and b) a symbol of wealth, a socially determined sign of a social reality. Their lack of a distinction between money as wealth in the form of an object and money as a symbol of wealth created a practice that excluded any other practice which could have made the social phenomenon of money create wealth through productive investment. The fact that this confusion reigned at a time when there was an increase in the possibility of producing more money had catastrophic results.

In the politics of Philip II, we recognize the same contradiction that we found in the economic actors. On the one hand Spain was to become rich and powerful through its supply of precious metals, so export was banned, and on the other hand the government itself contributed to the outflow of money in the form of outlay on imperialistic politics. The point is that these two components in economic policy were not seen to be related — they were never regarded as parts of a whole, so that they could be perceived as irreconcilable. The increasing poverty remained a mystery to individuals and the government.

In this example it is, as mentioned, money which comprises the field in which a contradictory practice is embodied. Somewhat surprisingly perhaps, it turns out that it can be quite illuminating to compare these ideas with Habermas's theory of money as a medium, which in turn was inspired by Parsons (Habermas 1987b, p. 265). According to this theory, money must basically be able to be disposed of in the following ways: a) It must be completely appropriable and disposable, i.e. it must circulate, b) it must be depositable, and c) there must be opportunities for investing it. The actors in a monetary economy can choose whether to accumulate or spend their money, save or invest it productively. Seen from this angle, the flaw in mercantilist policy appears on two levels: It is not possible for everyone to earn by saving if no-one invests, and it is not possible to perceive saving as investing, as money has no value in itself. In other words, not everyone can pursue the same strategy of saving, for then everyone loses.

Of course we cannot deny that this is rooted also in genuine ignorance, as was the case with the Chinese farmers, and not solely in an undisclosed form of dysfunctional and self-concealing social organization. In Sartre's perspective it seems, strangely enough, that all importance was attached to this last factor. Reification is not a phenomenon which afflicts man, in his view, and in this sense he adheres to his original concept of freedom, that man is always beyond himself in new projects and thus has no core from which he can be alienated. The *projects* however are reified because of the others and their influence. And social matter or materiality constitutes precisely the field where the various and uncoordinated projects come up against one another and become embedded. He therefore says of reification:

> It is not a metamorphosis of the individual into a thing, as is often supposed, but the necessity imposed by the structures of society on members of a social group, that they should live the fact that they belong to the group and, thereby, to a society as a whole, as a molecular statue (Sartre 1976, p. 176).

The significance of matter is then not only that it recreates this relationship between individuals, but also, to the extent that it constitutes a counter-practice, in the way in which "it links the meanings [of the individual projects, T. K.]" (p. 178). It is only then that what started out as spontaneity attains a reified character. It is then reasonable, as I have already mentioned above, to view the separation, the lack of coordination of the individual projects, as epistemologically more fundamental than the inertia of materiality. We must however remember that this is primarily an *epistemological* relationship of priorities; in the actual societies our history knows of, the inertia is omnipresent, even though it must be understood on the basis of deficient coordination. Man must orient himself in these kinds of surroundings, and for Sartre, machine technology, what he calls the iron-coal complex, seems to be the most developed example of this kind.

The phenomenon of anti-practice is therefore our confrontation with the others as mediated through machinery. However, the fact that Sartre can consistently claim that we are not really dealing with the reification of the worker, or his reduction to a nothing in relation to machinery, is justified by the fact that even if the anti-practice itself is active, it makes *demands* on the workers — the prescribed responses that these labouring individuals give are nevertheless entirely their own. It is through this adaptation, which occurs as an unintended result of earlier practices, that the practical-inert field constitutes itself as a field where the division between subjectivity and objectivity is annihilated. We do not become machines, but the machine is inside us. We are thus still people, as opposed to things, but we are the very "people of needs, practice and scarcity" (p. 185). This type of person is determined not least by alterity, the idea that the function of any individual can just as well be performed by someone else (assuming of course that they fulfil the given requirements). Furthermore the particular activity that I perform in a particular place is only mine in the sense that I am one like the others. This is not however only the alienation of atomization and the division of labour *between* individuals; since this is the way that everyone exists, then everyone is an other to himself.

The following passage provides unequivocal expression of this attitude in Sartre:

> In either case, material exigency, whether it is expressed through a machine-man or a human machine, comes to the machine through man to precisely the extent that it comes to man through the machine. Whether *in the machine*, as imperative expectation and as power, or *in man*, as mimicry [...], as action and coercive power, exigency is *always* both man as a practical agent and matter as a worked product in an indivisible symbiosis (p. 191).

The moment of *exigency* as inert, imposed finality makes it possible to conceive of the kind of negativity known as *objective contradiction* (p. 193).

Here too it is worth our while studying a concrete example, in order to see what is lurking in the shadows of the metaphors. Sartre refers to investigations of female skilled workers from the first phase of the semi-automatic machine (p. 233). As they worked they entertained erotic day-dreams. "But it was the machine in them which was dreaming of love", they could neither switch off nor concentrate fully, as this would interfere with the special mentality of supervised but not entirely conscious input, "an explosive mixture of unconsciousness and vigilance". (Men, it would appear, do not have erotic fantasies in this kind of situation because they associate eroticism with activity, not passivity and domination. Sartre supports these views with a reference to Simone de Beauvoir's distinction between the first and the second sex.) Once again the strength of materialistic phenomenology is demonstrated here, which I pointed out in connection with Leroi-Gourhan's concept of chains of operations (see Part II, chap. 1 above). Heidegger's analysis of to hand and at hand cannot accommodate *this* kind of pre-reflexive level, which rests on the idea that material entities will be or have already been embedded in the individual. I am not accusing Heidegger of considering a certain form of practice as genuine and theory as alienated. That would be an unforgivable vulgarization. I do believe however that I have found him wanting when it comes to his ability to discover how technological imperatives and thus social relations of power have become embedded in the individual on its seemingly most original levels.

A more conventional example is the analysis of anarcho-syndicalism (p. 235). Sartre finds its material fundament in the universal machine, a machine tool which could be adapted to many different functions. It could only be manned by a qualified skilled worker, thereby creating a hierarchy of qualifications and levels of pay among the workers, and developing a specific ideology among the privileged and the best paid.

The closest thing I have discovered in Sartre to a theory of the way in which a counter-practice arises in the inert field under our historical conditions, so that the requirements of the machine become the alienated person's relationship to himself, is his reference to the diachronic function of the practical-inert field, i.e. that it connects the past to the future.

> Inertia comes to him [the worker T.K.] from the fact that previous work has constituted in the machine *a future which cannot be transcended* in the form of exigency (that is to say, specifically, the way the machine is to be used and its ability, in definite conditions, to increase production by a definite proportion), and from the fact that this untranscendable future is actualized in all its urgency by present circumstances [...] Thus the *inertia of praxis*, as a new characteristic of it, removes none of its previous characteristics [...] But passivizing *annulment* modifies it from the future towards the past within the petrified framework of exigency: this is because the future to be realized is already fabricated as mechanical inertia *in the way in which past being is transcended.* [...] On the contrary, *precisely* because they have been worked and assembled by men, [...] the machine and the combination of exigencies contain the movement of transcendence in themselves (p. 235).

It is then still possible to claim that man is a creature that only exists through its projects. Man is and remains a creature that transcends itself, but these individual projects are now dependent on a future which is already partially structured through earlier choices, Sartre claims at this point in his philosophical development. I choose to regard this in the light of what I described as a fundament of materiality (see chap. 3): It links past and future by connecting our future possibilities to previously made choices.

# Part III
THE COMMUNICATIVE THEORY

# Introduction to Part III

The objective of the final section of this work is to discover how and to what extent insights from the theories I have discussed so far in this book can be integrated within the framework of the general social theory put forward by Jürgen Habermas. His theories can perhaps best be called the theory of communicative action. I have already justified my reasons for assigning his philosophy such a fundamental place in this book in the Introduction. In a nutshell, I believe Habermas's theory is the most promising approach within social theory today. It would be pointless to give an elaborate account of this opinion: Let me simply say that it appears the most interesting approach because it most explicitly takes into account and tries to unite the various motives that are inevitably included in a modern philosophy of society: The sense of the irreversibility of the process of modernization *combined with* a critical approach to society; the necessity of a universalist ethics *and* of the historical constitution of the subjects; of a rationalist philosophy of science *and* a position critical of scientism.

Nevertheless it may seem very strange to choose this particular theory of anti-scientism as a kind of template or formula for my evaluation of theories of technology itself. Both general and more specialized observations appear to be able to support this very view. Habermas has always been accused of abstracting from and idealizing concrete social relations, and pushing aside the specifically material aspects of society's reproduction of itself.

At first glance it is certainly implausible to say that Habermas underplays the role of technology in the theory of modern society. He first received recognition in the mid-sixties not least for theories which all, admittedly to varying degrees, emphasize the role of technology. I am thinking here of the theory of cognitive interests, one of which was the technical interest, which formed the basis of the natural sciences — the theory of technology and science as ideology and the distinction between labour and interaction. The question we ask ourselves here must be not if, but how exactly he approaches the problem of the role of technology.

However, Habermas's thinking underwent a change at the beginning of the seventies which I am going to call the linguistic turn in his philosophy. This entailed that the theory of the connection between technology and natural science had more or less been left hanging in mid-air, if not directly dropped altogether. The two other themes were based on intuitions which

Habermas has never given up, but which have been reformulated in a completely new way within the framework of a theory of communicative action and rationality, as in *Theorie des Kommunikativen Handelns,* here cited from the English version as Habermas 1987b and later works. More specifically we are probably justified in claiming that Habermas's theory relates in a strange and paradoxical way not only to the very theories I have discussed hitherto, but also to the strong intuitions we have that technology is in fact a fundamental feature of modern society. The theories I have discussed above, especially the concept of materiality in what I called the dialectical tradition, have all been developed within the confines of what Habermas calls the subject–object theory or the paradigm of consciousness in modern philosophy. Habermas uses this expression to designate the epistemological tradition which appeared in the period from Descartes to Kant. The theories of society put forward by Hegel and Marx are directly linked to this first formulation of the theory of consciousness in early modern societies. Materialistic phenomenology, as in Sartre, has its point of departure in Husserl's modern version of the theory of consciousness. Nonetheless all these philosophies of society, according to Habermas, remain inextricably connected to a model based on a single subject which in some way is confronted with an externally given object unknown to him. (Of course this is not to say that, for example, Husserl, in Habermas's opinion, does not have a theory of intersubjectivity, but rather that his theory is inadequate because the relationship between subjects is based on the model of the relationship between a subject and an object.)

In my opinion, Habermas's criticism of this theory is one of the most convincing arguments for taking the new and communicative basis he himself developed as a starting point. Furthermore it is of decisive significance for the development of his mature position, as found in *Theorie des Kommunikativen Handelns* and his later works, which will be central in the rest of this book. *At the same time,* however, this criticism also results in technology seemingly sliding out of the field of interest in his writing from the start of the 1980's and onwards in a way that is difficult to explain.

My discussion of Habermas will therefore have to attempt to solve a double problem: It must both prove that his theory includes a convincing critique of the insufficient preconditions of the earlier theories of technology *and* that it itself makes a positive contribution to this field. It will have to demonstrate that it is possible to discuss technology in a satisfactory manner within the framework of the communicative approach, and that it can incorporate the valuable results from the dialectical approach, albeit in a reconstructed form. The key concept here, as mentioned in the Introduction, is materiality.

The bulk of the emphasis will of course be placed on the latter, constructive task, and I will attempt to provide a solution to his theory of communicative reason by scrutinizing the texts, in particular Habermas 1987b, which illustrates how the role of technology in modern society despite everything has been made a theme in Habermas's own texts, and by demonstrating how they pave the way for and can successfully be supplemented with a treatment of this theme. It is interesting to note that one commentator, Axel Honneth, deems Habermas's interest in technology and theories of technocracy to be the fundamental theme in all his works, the treatment of which he regards with greater scepticism than I.

With regard to my own discussion of Habermas, it is necessary to underline once more that this book is intended to deal with technology, and has no pretensions whatsoever of providing a complete account of the authors mentioned. I would like to stress in particular that I will not be presenting any original views on Habermas's theory of rationality or his theory of language, with which I am assuming the reader is already acquainted. I will confine myself to touching upon those themes that are necessary in order to deal directly with the question of technology. I am hoping that my account will gain from this delimitation of its subject, despite the fact that it is unavoidable that certain themes and theories must be taken for granted.

I will introduce chapter 6 with an account of Habermas's theory of the cognitive interests, a subject that can be dealt with in relative isolation from the rest of his writing. The same cannot be said of his theories of technology as an ideology and the distinction between labour and interaction, which are central concepts in his work from the 1960's. Here we come across intuitions that later reappear within the framework of the theory of communicative action, but then on the basis of an entirely different theoretical context. Nevertheless I believe that the original theories, as well as having been of great significance when they were first presented, still have elements of great universal interest today. I will outline these theories from the 1960's in chapter 7 and deal with the foundations for the new theory in chapter 8, which briefly sketches the critique of the subject–object philosophy. What I have called the constructive part of my thesis appears in chapters 9 and 10, which together comprise a review of the possibilities for discussing technology on the basis of the theory of communicative action, as this theory has been developed in Habermas's texts from 1981 on.

# 6 The Theory of the Technical Cognitive Interest

In the following discussion of the relationship between on the one hand science and on the other technology and instrumental action in Habermas's early works, I will be concentrating exclusively on the technical cognitive interest within the trichotomous theory of the cognitive interests, which was developed in the book *Erkenntnis und Interesse (Knowledge and Human Interests)* Habermas 1987a, and in the essay of the same name in this book (pp. 301-317). Within this set of problems I am primarily interested in the relationship between instrumental action and scientific research, i.e. a reflection on technology from the angle of the theory of science.

By way of an introduction it might be profitable to clarify two themes from the general debate about the early part of Habermas's philosophy. The first concerns the general architectonics of Habermas's theory. My main interest is not the multitude of accusations made against Habermas for falling back on the approach of neo-Kantianism, in his attempt to demonstrate how a certain number of research logics determine their respective fields of research, the types of question they ask and their forms of objectivity. This would take us far beyond the scope of this project. I am thinking of a different problem here, namely the internal relationship *between* the three different cognitive interests. Taking an article by Tronn Overend (Overend 1978) as our starting point, we can formulate the question as follows:[16] Which form of knowledge and thus which cognitive interest underlies our knowledge of the three cognitive interests? The only possible answer is the liberating interest. It seems obvious that the cognitive interests and their basis in natural history are disclosed and discovered by means of the concept of self-reflection which Habermas borrows from German idealism and which is most intimately linked with the liberating interest.

> The mind can become aware of this natural basis reflexively. [...] But the mind can always reflect back upon the interest structure that joins subject and object a priori: this is reserved to self-reflection. If the latter cancel out interest, it can to a certain extent make up for it (Habermas 1987a, p. 312).

---

[16] I would like to stress that I agree with Overend only on this one point. His critique of the consensus theory of truth is unacceptable in my view.

One could of course argue that we have now reached a meta-level, but it is most reasonable to say that it is the interest in liberation that acts as a meta-level, rather than claiming that a fourth interest has been introduced. Two questions nevertheless arise:

1) If the doctrine of the cognitive interests is determined by the liberating interest underlying the social sciences, then the autonomy of the other sciences is also threatened. In particular it seems that the logic of the natural sciences would be reduced to a criticism of the ideology behind the forms of thought applied within the natural sciences and that this in turn leads to a form of idealism and relativism. As we shall see, these kinds of tendencies are not without foundation in Habermas's earliest publications, but appear to be irreconcilable with the theory of the three different forms of science.

2) The same argument can be reformulated in slightly different terms. If the theory of, for example, the technical cognitive interest itself is a product of an interest, it is then difficult to perceive it as anything other than some form of empirical knowledge. It is clear straightaway that the cognitive interests, according to Habermas himself, have a quasi-transcendental or an empirical-transcendental status. This is diffuse enough, but the point is that this criticism affects *even* the weak form of transcendental status which Habermas asserts.

> As long as these interests of knowledge are identified and analyzed by way of a reflection on the logic of inquiry that structures the natural and the human sciences, they can claim a "transcendental" status: However, as soon as they are understood in terms of an anthropology of knowledge as results of natural history, they have an "empirical" status (Habermas 1971, p. 21).

It is this transcendental status itself, as an a priori precondition for a field of objects, that is lost if the cognitive interests are not developed in parallel, but rather the two of them depend on the third.

The *second* problem concerns for the most part the technical interest, and touches upon the relationship between nature and the subject within this interest. Here too Habermas's formulations are vague, but the problem arises on the basis of theses one and two in the article "*Erkenntnis und Interesse*", which at the same time claim that the subject has emerged by means of a natural history which it transcends.

This has led to writers like McCarthy and Skirbekk enquiring whether this connection can actually be formulated consistently. How can the subject be brought forth from and bound to nature and at the same time be the entity that constitutes this nature? (Cf. Skirbekk 1993, p. 172 and McCarthy 1978, p. 115). In one version this alleged paradox seems to me to be purely verbal.

First of all we can distinguish between two meanings of the verb "to constitute": We are constituted (1) by natural history through the processes of evolution. In this way a species appears which is actually forced to relate to nature through labour, i.e. instrumentally. And we constitute (2) nature in the sense that we have access to it only by means of a level of instrumental action. This need not result in a paradox if we interpret "constitute (2)" in a sufficiently weak sense as implying that our apprehension and experience of external, consciousness-independent objects that affect us causally is associated with a function circuit of action and registration of the "responses" of the objects to our actions. At this stage it is not problematic to operate with two meanings of the verb to constitute, as long as we understand that constitute (2) is logically primary to constitute (1). Our knowledge of processes linked to constitute (1) is a purely empirical knowledge of the nature to which constituting (2) grants us access. (At this point we must remember that the idea that science has anything whatsoever to do with instrumental action is of course still pure postulation.) What we need here is the concept of emergence, that at some point in the biological development a new level emerges, with entities that did not exist on lower levels. In our case this will mean that we reach that point where we come across human subjects, equipped with abilities that did not exist on previous levels.

So in the first instance I regard the idea that there is something odd about the nature that constitutes us — "nature coming from behind", in Skirbekk's words — as merely a pseudo-problem. All we need is a general evolutionist theory, since we are dealing with the same nature, albeit conceptualized on vastly different levels of sophisticated systems of concepts.

McCarthy and Skirbekk, however, have more up their sleeves. Since Habermas to a certain degree builds on the philosophy of consciousness as developed by German idealism, in which we primarily have access to ourselves through a specific operation called reflection, he has taken over not only pretensions of aprioricity from Kant's epistemology but also the conception of a nature in itself. McCarthy demonstrates that the connection between a concept of nature in itself and a constitution of nature, and between the idea of a nature in itself and the constituting human species, which is a product of this same nature, is highly problematic. With the following I am hoping merely to propose one consistent solution, as opposed to trying to salvage all of Habermas's contentions. I find that McCarthy convincingly proves that Habermas stumbled into these problems because of his view of the theory of evolution. McCarthy discusses this problem in what he calls an epistemological, a subjective and a natural-historical context (McCarthy, p. 115), but it appears that the problems reach their peak for him in the last respect:

...in this context our dilemma takes the following form, how can the subject that transcendentally constitutes nature be at the same time the result of a natural process? (p. 121).

He ascribes to Habermas the following conception:

"Objective nature for us" is constituted, the result of a "synthesis" subject to an interest in technical ontology. As such it obviously cannot be prior to the human world (p. 123).

Had "constituted" actually meant "processed", then McCarthy would of course have been right. But if we understand "constituted" in sense (2) above, then we are merely saying that our concepts of nature are associated with an original level of instrumental action. They indicate to us the conditions under which we can understand objects and processes, when they occur. The fact that Kant conceived of time as an a priori form of intuition for mankind only, implies only that nature for us always exists in time, not that it emerged at the same time as us in time.

This interpretation, as I mentioned above, has no pretensions of salvaging all of Habermas's theories, in the sense that it will make everything he said consistent and clarify all his unclear points. I should then perhaps give an account of what I believe I have *not* been able to establish in this description.

In particular I am presupposing, despite Habermas's own words, that evolutionist theories are purely empirical theories of natural science, independent of a concept of reflection. All we need here, as already mentioned, is a concept of *emergence*, i.e. that new phenomena emerge on new levels of biological development, that with the human species (or perhaps I should say with early humans) the abilities emerge that are a necessary precondition for the formation of societies — the abilities both to labour and to interact.

Nor is it by any means clear that we in this way attain one and the same concept of nature in the three cognitive interests. In later works this does not appear to worry Habermas unduly. He merely ascertains that a variety of concepts of nature exist here (Habermas 1985c, p. 212).

Furthermore the transcendental status of the subject remains unclear in the cognitive interests, even if it cannot be charged exactly with McCarthy's incriminations. The problem is rather that with a concept of interest which must range from the necessity of biological survival to theoretical knowledge, Habermas is on the verge of falling prey to a form of naturalistic reduction of his concept of knowledge. It is not so much a metaphysics of nature, as McCarthy fears, that emerges here (McCarthy 1978, p. 124), as a classical metaphysics of the subject typical of the seventeenth and eighteenth centuries. (For an account of this entire problem area, see Blumenberg 1976 and the other contributions in this volume, especially Ebeling and Henrich.)

Perhaps it is a symptom that this problem area seems to make its way into a post-Kantian philosophy more easily than the metaphysics of nature (cf. Sommer 1977).

The crux of the matter is of course the transcendental element in the theory of the cognitive interests. I have not tried to reconcile the empirical and the transcendental in Habermas, but I believe I have found a way of avoiding the conflict by severely reducing the transcendental, in favour of the purely empirical. All that remains is that we can talk of the preconditions for science as a priori, in the sense of preceding, of constituting the necessary conditions for the given fields of objects, but not in the sense of being necessary conditions or forms of knowledge in themselves.

Nonetheless McCarthy could accuse me of merely shifting the problem over from the distinction between man and nature to the point where we must distinguish between natural and social evolution, the point where the emergent factors appear. And he would undoubtedly be right. My defence, however, is that we at this point are faced with a constellation that is not paradoxical, but which can be dealt with on the purely empirical level.

Finally I would also like to point out that my interpretation is the one that is most concordant with Habermas's later development. I shall only refer to his arguments in order to avoid the label transcendental pragmatism (Habermas 1979, p. 21) and the introduction of the theory of a purely empirical-reconstructive theory of human capacities with the criticism of the concept of reflection that it implies (see Habermas 1987a, Postscript). Finally, in Habermas, we read:

> Naturally, I am enough of a materialist to take as my starting point that Kant is right only to the extent that his statements are compatible with Darwin. I have never had any doubts about the primacy of natural history over the history of the human species (Habermas 1985c, p. 213).

Having clarified these points, I would now like to tackle my main topic. For the sake of clarity I would just like to remind the reader that Habermas's theory of the cognitive interests, which was primarily developed in Habermas 1987a, is itself a result of a path of development which is marked out by his first publications on the philosophy of science.

The first phase of his authorship on the philosophy of science was closely connected to what was commonly known as the positivism debate in the 1960's, a debate which caused a great stir, even in provincial Norway. The debate was rooted in the controversy between Adorno and Popper at the German Congress of Sociology in 1963, and revolved around Popper's philosophy of science.

It was in fact in this debate that Habermas developed his views through, at least in intention, an immanent criticism of first and foremost Popper's philosophy of science. In retrospect we can focus our attention on two issues in this debate. The first is the more general claim that the theory of the value neutrality of science concealed an association with technical applications of scientific knowledge and therefore a technical interest. The second point of contention was the more specific thesis that this association was expressed by a link between the growth and justification of scientific theories and a specific historical level in the social process of labour which European societies had reached in the early modern era.

Together these two assertions were supposed to demonstrate that the association between science and technology lay in the predicative character of scientific knowledge. The common feature of science and technology was supposed to consist in the fact that both produced preconditions for confirming our expectations about the course of objective processes. Based on this, Habermas concluded that natural science built on an interest in control of objectivated natural processes. The requirement of value neutrality is consequently threatened:

> It would indeed be endangered if, for the modern sciences, through an immanent critique, a connection were demonstrated with the social labour process, a connection which penetrates the innermost structures of the theory itself and determines what shall empirically possess validity (Adorno et al. 1976, p. 156).

In these early texts, which were clearly influenced by Adorno, Habermas stuck his neck out by maintaining that modern natural science was directly dependent on the production processes of early capitalism. Habermas claims that Galileo's natural science grew out of a form of technical control which corresponded to that which was found in the type of technology that Marx called manufacture, which was based on analysis and subdivision of the labour process into its simple and constitutive components (see chap. 5 above).

What is striking here is that, despite all the differences on general levels of philosophy, we find again and again that Heidegger and the Frankfurt School are in fact in agreement on specific theses as regards their view of technology: Habermas and Adorno of course perceived Heidegger's ontologizing of history, his interpretation of the unfolding of technology as *Seinsgeschichte*, as an anathema. But the idea of anchoring all forms of reflection, all intellectual expressions, in concrete historical conditions, unintentionally leads to the same end. On the basis of the opinion I have outlined, science is, both as a phenomenon specific to the historical epoque and in its structure, quite dependent on technology. The traditional view of technology as applied knowledge thus loses its foundations; instead it is theory that is applied practice, as Heidegger had already claimed. (For a more precise formulation of the individual theses, see Heidegger 1977. There are several reasons why Habermas may have chosen to retreat from this somewhat dubious position. One is of course that he may have recognized an essential distinction between two historical theses. The first is that early modern science presupposed its contemporary technology in order to be able to acquire adequate equipment for its experiments, at the same time as mechanical devices may have been of heuristic relevance to the development of individual hypotheses. The second is the attempt to justify in terms of the sociology of knowledge that which Heidegger arrogantly only assumes, namely that the structure of knowledge is in some way dependent on a historical element in the production process. The first is a possibly true empirical hypothesis. The second is a speculation that is methodologically unsound. It is interesting that Habermas should turn to the work of Franz Borkenaus for support. In the early days of the Frankfurt School Henrik Grossman had already proved that this was a position which was permeated with non sequitur argumentation, false chronology and the weakest and most imaginative "analogies" and "likenesses".

This kind of exaggerated contextualization of the structure of natural science can of course easily result in a form of relativism. Questions of empirical validity may become directly dependent on technical success. Since the issue of whether a theory has been supported or weakened in an experiment is linked to the criteria for success or fiasco in technical operations, the answer to this question is in turn indirectly dependent on the social criteria for what comprises a technical success or fiasco, the social conventions of certain groups (p. 39). To avoid any potential misunderstandings, let me make it quite clear that Habermas did not, at this point or later, ever claim that the question of truth could be determined by interests. It is the forms of the questions and of the validations which are perceived as being interest-related. What we are interested in discovering is how the processes proceed; the interest cannot determine the answers to scientific questions. (The interpretation that Habermas believed that the driving force behind the natural sciences was a conscious motive to acquire applicable knowledge is such a vulgar objection that I see no grounds for going into it here.)

Habermas has himself later explicitly corrected these theses, but even after his theory has been cleansed of these elements from the bestiary of the sociology of knowledge, there can hardly be any doubt that he confuses and overplays two otherwise interesting theses in a possibly immanent criticism of Popper. Popper had contended that both the question of whether an experiment is successful or not, and whether a basic statement should be accepted or not, are matters that must be decided by a research collective. One could of course claim that this admission, in contrast to classical positivism, paves the way for a hermeneutic dimension in the natural sciences as well; the results of research must be *interpreted*. Scientific progress is thus continuously connected to reinterpretations of the available data, and not only to the procurement of new data, a fact that will be of significance later in this section. (As an aside I would like to point out that it is not implausible to claim that this point is made more clearly in Popper's standard version of the theory of science than in Habermas's, who at this point in time had a tendency to accept a relatively primitive, positivist view of the natural sciences, in order to then find a hidden dimension in this version.) It might also be an interesting thesis that Popper in his theory of observation statements unintentionally implies a consensus theory of truth. But of course none of this implies that form of relativism to which the coupling of test results to criteria for technological success comes dangerously close.

Judging by his later texts it also appears that Habermas has quietly dropped a thesis which appears in his earlier works, and which stated that the main connection between technology and science was situated in a role which the technological application of knowledge plays for the actual testing of our scientific theories. It would appear that Habermas is visited by the idea that the verification, which can never be achieved in the scientific process itself, is attained externally in and through the technical use of our knowledge. If we avoid presupposing what Popper assumes, namely that verification, if this *contra impossibilem* were at all possible, could only occur in the relationship between data and theories, then we could put forward the thesis that verification takes place in what Habermas calls result-controlled activity, and which designates a form of activity where our implicit and action-related presuppositions about objects and processes are continuously put to the test:

> [If we]...take technique in its widest sense seriously as a socially institutionalized regulatory system which, in accordance with its methodical meaning, is designed to be technically utilizable, one can conceive of another form of verification. The latter is exempt from Popper's objections and concurs, in fact, with our pre-scientific experiences. All the assumptions, then, are empirically true which can guide feedback-regulated action without having been previously rendered problematic through errors experimentally striven for (Adorno et al. 1976, p. 208).

Even though this thesis still seems rather unclear, Habermas does not make the mistake that adherents of Popper are so pleased to discover in pragmatic philosophies of science: He does not claim that theories are tools or means, but that theories of natural science formulate knowledge in such a way that the setting up of theories and testing is only meaningful insofar as this knowledge itself can be applied technologically. It is however possible to ascribe to him another mistake that Popper finds in pragmatic thought, namely that technical success — the constant application of knowledge for technical purposes — is a guarantee for the validity of our knowledge, even a form of theoretical confirmation. The point is that it is perfectly possible to achieve satisfactory results on the basis of theoretical presuppositions that we know to be untrue, such as when we continue to use an antiquated physics in engineering science (Popper 1969, p. 56). Therefore, even if technological success does provide us with a confirmation of prescientific opinions, it does not necessarily also provide us with something analogous to a confirmation of scientific theories in the methodological sense. Scientific confirmation simply does not take the form of a range of cases where what we thought worked in a technical context appears to have done so once more. The decisive point is that what counts as a confirmation in a scientific test must be judged contextually — contextually here of course in a statistical, as opposed to cultural, sense. It is the confirmation of the predictions that were deduced logically from a hypothesis, and which otherwise would be unexpected and surprising, which can be said to confirm the hypothesis. This, in my opinion, is the significant outcome of the debates on what constitutes a confirming instance, for example in connection with the paradox of the ravens. (For a survey of this debate, see Lambert & Brittan 1987, chap. iii, esp. parts 3-5.)

Habermas got himself out of this impasse by means of decontextualizing the concept of technology and labour. Labour and technique are considered invariant and constant. In the final version of the theory of the cognitive interests we must be justified in saying that they play the role to which we would like to be able to attribute transcendental structures; they are at any rate a priori in the sense that they are given to us human beings as unavoidable, actual preconditions as we saw above. For that matter we could join Raymond Geuss in saying that in the Frankfurt School there is both a contextualistic and a transcendental paradigm, represented by Adorno and Habermas respectively (Geuss 1981, p. 63). But this applies only to the situation at the end of the sixties. Habermas's later version of his theory makes it unreasonable, as he himself explicitly states, (see the reference to Habermas 1979 above) to call his theory transcendental. Geuss overlooks this because he combines a critique of a transcendental theory of the cognitive interests with a critique of a universalist ethics; two questions which are obviously not in the least related. The answers will at any rate not depend on the same set of arguments.

One point which concerns the theory of science seems to be fairly clear: The theory of the cognitive interests functions as a part of an anti-objectivist position. Prior to the various types of science there is an attitude that determines and delimits a fixed field of research and thus provides each type of science with a concept of its object(s) and its own form of objectivity. Nevertheless this does not yet tell us anything about how the relationship between the scientific construction of theories and instrumental action will appear on the basis of this theory.

However, before I turn to this point, which comprises my main interest for a closer critical examination, I would like to indicate a first positive result of this transcendental, or at least approximately ahistorical, version of the relationship between technique and science. It is expressed in the criticism of the exaggerated politicizing of the technological rationality that was found in the first generation of the Frankfurt School, and in particular in the writing of Marcuse.

Marcuse had first identified Weber's concept of purposive-rationality with technological rationality.[17] In the next round he could then follow the basic structure of Weber's theory of the rationalization of the Western societies by understanding these societies as permeated and determined at all levels by technological rationality, and at the same time insist that we are confronted here with a perverted and false ("one-dimensional") form of rationality. (In this way he carried out Lukacs's transformation of Marxism from a theory of materiality and forms of ownership to a theory of rationality and rationalization.) Modern society is to equally high degrees determined by and permeated by technological rationality. The flaw in this theory is not so much that it indicates that a programme for political change will have to query the manner in which technological choices are made and technology is controlled today. Its weakness is rather the opposite: That political change must now primarily take the form of, and indeed is only possible as a form of an unspecified but qualitative technological change which leads to a new technology, a new form of natural science.

It is precisely here that the transcendental, or at least the superhistorical, version of the cognitive interests have political significance. By linking technology and science to an unalterable level of purposive-rational contact with physical objects, Habermas manages to circumvent a pitfall: The idea of the

---

[17] For the following account of Marcuse, see in particular "Some Social Implications of Modern Technology" (Marcuse 1941), where his debt to Weber is most obvious. The general qualities of Marcuses theory that I am accentuating here, are, in my opinion, equally present in One Dimensional Man (Marcuse 1964).

transcendence of modern society is no longer connected to the purely speculative conception of a different technology and a different science to those we know[18] (cf. *Technik und Wissenschaft als "Ideologie"*, in Habermas 1971, p. 85). For a more detailed account of Marcuse see Krogh 1991a, in particular chap. 7.

In the following I shall be concentrating on the impact of the connection between science and instrumental action in Habermas's theory on the theory of science. One objection can be raised immediately if we regard Habermas's point of view primarily as a theory of *technology*. At this point Habermas relies directly on Gehlen and his theory about the relief of organs (Habermas's *Technik und Wissenschaft als "Ideologie"*, in Habermas 1971a, p. 86, see also Part I, chap. 2 above). I am therefore the first to admit that the thesis of the technical cognitive interest is hardly of much value to us if it is taken to be a theory of technology seen in isolation. This is however hardly a sensible approach. Habermas does not need a complete theory of the economic, cultural and institutional preconditions for technological change, but only to isolate one historical invariant level in our relationship to nature. There is, however, another more pressing issue: Is a theory which primarily regards technology as relief particularly well-suited for grasping the essence of the technical operations that are associated with scientific experimentation and the use of laboratory equipment? The aim of scientific experimentation is after all to discover something *new*, test contrafactual hypotheses or give the grounds for what is counter-intuitive. We must at any rate always be open to the unexpected, rather than presupposing the expected. This objection is of course closely related to the objection I made above. We will have opportunity later to go into it in more detail in connection with the relationship between experimentation and instrumental activity.

How does the relationship between technology and natural science seem now on the basis of this new version of the technical cognitive interest in natural science?

---

[18] This thesis demands a more precise formulation. The unsurpassable element of technology that renders another technology impossible is grounded in its foundation in instrumental action. This does not necessarily imply that new techniques cannot lead to unforeseeable cultural changes. As far as the idea of a new science is concerned, it may perhaps seem reckless to deny the possibility of it, bearing in mind the great revolutions that have taken place in the history of science. My rejection of the expression a "new" or "different" science does not entail a limitation of the possible content of future theories. I am promoting a different and probably not uncontroversial notion. We cannot associate anything with a concept of a science that would not have an objectivating approach to natural phenomena and which we at the same time are supposed to be able to recognize as a continuation of "our" science.

In the essays in *Technik und Wissenschaft als "Ideologie"* (Habermas 1971a and 1987a), it appears that this relationship can be summarized in two theses, which refer to the structure of the theories and the procurement of the data that can be used to determine the truth value of these theories, respectively. The first thesis hooks on to the fact that hypotheses in empirical natural sciences must be formulated in such a way that we can deduce testable implications from them:

> Empirical-analytic knowledge is thus possible predictive knowledge. However, the *meaning* of such predictions, that is their technical exploitability, is established only by the rules according to which we apply theories to reality (Habermas 1987a, p. 308).

With a certain amount of goodwill we can also find this thesis in Max Horkheimer (Horkheimer 1937). This quotation also demonstrates that Habermas's main point here is the location of the hypotheses as antecedents in a deductive form of the type *modus ponens*, a fact that lends the deduction of methodological preconditions the same logical form as the deduction of prognoses, and not the semantic form of the hypotheses themselves, as Dagfinn Føllesdal asserts with reference to Horkheimer (Føllesdal, Walløe, Elster 1986, p. 120).

We have already come across the second thesis: That the data we use to evaluate the validity of the hypotheses is procured in ways that seem analogous to instrumental intercourse with natural objects.

> We can say that facts and the relations between them are apprehended descriptively. But this way of talking must not conceal that as such the facts relevant to the empirical sciences are first constituted through an a priori organization of our experience in the behavioural system of instrumental action (Habermas 1987a, p. 308).

Elsewhere this is formulated slightly differently. The type of test conditions is:

> [...] imitation of the control of the results of action which is built naturally into systems of societal labour (Adorno et al. 1976, p. 208).

Habermas thus believes he can draw the conclusion that:

> Taken together, these two factors, that is the logical structure of admissible systems of propositions and the type of conditions for corroboration suggest that theories of the empirical sciences disclose reality subject to the constitutive interest in the possible securing and expansion, through information, of feedback-monitored action (Habermas 1987a, p. 308).

This interpretation of the connection between technology and science is however still plagued by some obscurity. To begin with we can note that it remains unclear exactly *how* Habermas envisages the relationship between the two theses cited above. Are they two reciprocally independent arguments which both support the same thesis, that natural science builds on a technical cognitive interest? Or is the necessity of such an interest only apparent to us in the interaction between the organization of the system of hypotheses and the generation of data in the experiment situation which itself is analogous to technology?

This lack of clarity becomes more acute if we take yet another matter into consideration: It remains unclear how and on what level the second thesis is to play a role in the philosophy of science. It seems that Habermas has combined two different interpretations in this thesis. The first version, as I shall call it, of thesis two, means insisting on the analogy between labour processes (no doubt we are dealing with labour processes here that build on the machine technology of modern societies and those forms of technology that have replaced it — in short we are dealing with mass production) and experimental checking of hypotheses through attempts at reproduction of the results of experiments under identical test conditions. *Or* we could imagine that what in the terms of Habermas's later vocabulary could be called the objectivating attitude to nature, as a field of physical objects and processes, is only opened to us on the basis of a *previous* instrumental contact with natural objects. I will call this interpretation the second version of thesis two. In this sort of contact with nature we encounter it as objects whose reactions we can and must both control and predict. The basic features of the form of action that are expressed in this form of instrumental intercourse with nature can advantageously be thought of as historically invariant and linked to the self-preservation of mankind.[19]

---

[19] As McCarthy points out, Habermas's terminology for distinguishing between various types of action is rather imprecise (McCarthy 1978, p. 28). Habermas claims that by the term purposive-rational activity or labour he understands instrumental action *and/or* rational choice which are then called strategic action. Instrumental action in turn originates in prognostic knowledge of the outcome of observable processes. McCarthy is undoubtedly correct in his observation that this is not a separate *type* of action, but an element of strategic action, that information which renders possible a choice of a means to an end. I will return to this problem in detail in the next chapter. This might, however, be a good opportunity to deal with another issue on which I side with Habermas against McCarthy (for the following, see McCarthy 1978, p. 67). There is of course much confusion about the level, but it is wrong to try to argue against the superhistorical position Habermas tries to attribute to the instrumental approach to nature, by referring to all the non-instrumental relations to nature we can trace in the history of mankind. First of all, it seems reasonable to ascribe to all societies a common level which consists of the perception of and contact with physical objects (cf. Horton 1982). In most societies with which we are familiar, this

This distinction may seem fairly irrelevant to some readers of Habermas. I beg to differ, however, since later developments of Habermas's theory went in the direction of developing a relatively differentiated form of pragmatic philosophy of science, which drew an ever sharper line between a primary level of instrumental intercourse with nature and the secondary level where the specific theoretical problems are formulated and solved.

This development takes place by means of a thorough reception of Pierce's version of pragmatism (in Habermas 1987a), a movement that he himself describes as a development away from Dewey and towards Pierce, a form of pragmatism which also includes a healthy infusion of scientific realism. The destination of his development and the potential inherent in his final position is most clearly expressed in the postscript to this book, which he wrote five years later (in 1973). I am therefore going to concentrate on this work, but would also like to point out that in 1968 Habermas already perceives the relationship between instrumental action and science as a learning process which starts in a pre-reflected phase and moves to a reflected phase, and where the field for instrumental action now exclusively makes up the first pre-reflected and pre-scientific level.

> The process of inquiry, though, satisfies three additional conditions: (1) It isolates the learning process from the life process. Therefore the performance of operations is reduced to selective feedback controls. (2) It guarantees precision and intersubjective reliability. Therefore action assumes the abstract form of experiment mediated by measurement procedures. (3) It systematizes the progression of knowledge (Habermas 1987a, p. 124).

This already serves as a clear indication that it is the second version of thesis two that will triumph. This is also precisely the result of the development of his later theoretical work.

In 1973 Habermas had developed his consensus theory of truth. This theory allows him to develop a distinction which was not clear in the original

---

approach is inextricably embedded in other forms of rationality, rather than forming the basis of the prominent interpretations of the cosmos and man's surroundings. This does not mean, however, that it was not present. The Middle Ages could not live on religion, Marx wrote. Only relativistic anthropologies believe that man can live on magic. (This may nevertheless be a possible occupation in sophisticated cultures and in modern times.)

There is however one more problem here: Is it reasonable to believe that instrumental action forms the foundations of all natural science, e.g. biology? Gunnar Skirbekk in particular looks at the question of whether Habermas's concept of science is not too tailor-made for physics (Skirbekk 1993, p. 146). Biology and physics, both via reduction and as a coherent methodological unity, seem to be so united today, that I would maintain that this is not a major problem, assuming a *loose* connection between science and instrumental action.

text or in his previous manuscripts. I am thinking here of the distinction between objectivity and truth, and the corresponding distinction between the objectivity of experience, in a relatively conventional Kantian sense, and the pretensions of validity in theoretical statements. In the first case we are interested in the constitution of those forms of objects to which empirical statements refer, what Habermas calls the categorial meaning in these kinds of statements. Habermas refers to categorial distinctions such as those between things and processes, between people and their expressions as examples of the various forms of categorial meaning (Habermas 1987a, p. 360). It is on this level of *categorial* meaning of the statements that the objectivity of the experience is established. Their *discursive* meaning is linked to the pretensions of truth which can only be redeemed through argumentative grounding. Habermas thus distinguishes systematically between the conditions for the objectivity of our experience and the conditions for whether an assertion can be true or false.

> Since experience claims to be objective there is a possibility of error or deception. In cases like these, an opinion expressing (alleged) experience is 'merely subjective'. Yet, this *objectivity* of stated experience is not the same as the truth of the statement (p. 363).

> While the conditions for experiencing something objectively can be clarified in a *theory of the constitution of objects,* the conditions of argumentative reasoning can be clarified in a *theory of truth* designed as a logic of discourse. The two are not the same, although they are connected with each other via the structures of linguistic intersubjectivity (p. 365).

It is obvious then that Habermas is not claiming that the level of the constitution of the object is prelinguistic. It is here that we express those of our experiences that may be based on sensory deceptions and mistakes, such as if we believe that a sunlit spot on a darker background is a rock. If, however, we are searching for a theory of the "redemption" of those claims to truth which are implicit in the statement "This is a rock" we are transferred on to another level, according to Habermas. This is the level of *discourses* where the merely implicit pretensions of truth from our prescientific level of experience are made explicit and are put to the test. Since he is defending a consensus theory of truth and explicitly treats it as a contender to, for example, classical correspondence theories, we must to require that a theory is developed here about the conditions we must set for the truth of the proposition "This is a rock", which do not build on a presumptive concordance between the proposition and the facts.

The consensus theory of truth may as yet not have been fully developed by Habermas. It also seems clear, as Gunnar Skirbekk claims (Skirbekk

1982), and Habermas himself has vaguely confessed, that we have to find a place where references to sensory evidence can play a role in a discursive redemption of the claims to, for example, an observation statement. It would take us beyond the realms of my project to go into this debate in detail here. It nevertheless seems to me that this distinction between the constitution of objects and discourses that Habermas has introduced is so fruitful that there are grounds for further study of the significance that the distinction between the objectivity of experience and claims to truth may have for issues of interest to the philosophy of science. Furthermore, this set of problems touches directly on the relationship between technology and science. (Perhaps the idea of such a distinction between different levels may even function without presupposing a problematic consensus theory of truth.)

It goes without saying that scientific theories belong on the level of discourse. What we call objectivity in a scientific context is associated with the pretensions of validity of theoretical statements. This is objectivity in a different sense to what we mean when we talk about the objectivity of experience. Generally speaking, we can say that in this way the gap between the theoretical level and the level where the forms of the objects are constituted through instrumental action is widened. The decisive point for *theories of science* is that this forces a fundamental change in our view of the status of scientific experimentation, as I hinted above. While the function of experimentation was previously, i.e. in the first version of thesis two from 1965, conceived of as the main evidence for the instrumental character of science, Habermas now realizes, having been inspired by Hans Schnädelbach's criticism (Schnädelbach 1972), that experimentation cannot be interpreted as an example of instrumental activity in the prescientific sense. Schnädelbach distinguishes within the concept of instrumental between technical and experimental activity, where technical action almost stands for instrumental activity in the sense in which Habermas uses it.

Schnädelbach's point, assuming I have understood him correctly, does not in fact deny that experimentation consists of an application of the most complex and theory-dependent technology. In contexts of technological action, however, our action-related basic understandings, our "beliefs" in Pierce's definition of the word, are not questioned. Experimental action does not distinguish itself from technological situations through our necessarily discarding *these* presuppositions. On the contrary, we continue to perceive them as unproblematic, since the experiments themselves presuppose the use of technological equipment, in order to be able later to ask *further* questions about theoretical assumptions. It is only when the experiment goes wrong that the technology itself becomes the target of research in the search for possible flaws in the limiting conditions. Needless to say, the point here is that experimentation, especially in the form of formalized measuring operations, is an activity that is completely permeated by theoretical categories and determined by objectives that only emerge on the discourse level, such as, for example, the testing of universal hypotheses, counterfactual statements, etc., ultimately what Habermas calls "the most unnatural idea of truth".

> Experimental action does not lie on the same level as the instrumental action of naive or scientized practice. In its function of producing data, which is always gathered with a view to testing hypothetical validity-claims, experimental action is related to discourse from the start (Habermas 1982a, p. 275).

At this point it would appear that we have come a long way from the original idea that science is associated with self-preservation through its foundation in instrumental action. Even if instrumental action is necessary in order to maintain society, this is not the case with most experimental activity (see Stockmann 1978, p. 27).

The further consequences of this interpretation for the theory of science have been best developed in a thorough commentary by Mary Hesse (Hesse 1982), on which I will be basing my discussion below. The pragmatic or instrumental character of Habermas's theory of science is further weakened in this final draft. But the significance of empirical terms and statements is still established, according to this theory, on a level determined by instrumental activiy. We can therefore speak of a *combination* of a pragmatic or instrumental theory of the significance of empirical terms and statements and a theory of truth based on a discourse theory. Since truth is exclusively linked to the discourse level, i.e. to the grounding of theories, there are no remains of a pragmatic theory of *truth* in Habermas's relatively original conception.

The pragmatic level thus ensures us of observation statements that are relatively independent of the level of the theory. Or at least they remain constant, irrespective of our opinion about the mutually exclusive theories, as they are primarily anchored in an instrumental activity that can be reconciled with various theories. In this way we can be certain that denotative terms, despite the competition between theories, refer to identical objects in the world. As a consequence total variance of meaning is excluded, as far as empirical terms are concerned or such a variance is at least now independent of the rival theories and it is this that is decisive here. This in turn then allows comparative interpretations and confrontations between rival theories. Theoretical progress can thus be understood as various interpretations of the same data, which have been formulated in the empirical statements on the primary level. As we see, this theory is in turn dependent on the notion that truth is *only* questioned on the theoretical level in the way accounted for in a discourse theory. The function of the pragmatic theory is precisely to separate the level of the denotative function of empirical statements from questions of truth on the theoretical level, and thus *in turn allow* the association of these kinds of questions with a level that permits inter-theoretical comparison.

We can take Hesse's deliberations a step further. The strength of this theory and the specific, albeit somewhat restricted, use it makes of instrumental action in science, lies in the fact that it circumnavigates the relativisitic implications of Kuhn's theory. When Kuhn can claim that paradigms are incompatible because they do not refer to the same data, he is presupposing, as far as I can see, a holisitic theory of meaning, in the sense that the meaning of the empirical terms a paradigm recognizes is tied up with the meaning of the theoretical terms. The strength of Habermas's theory, although it is not the only one to do so, is that it illustrates a way to avoid such a completely unrealistic presupposition.

At this point we must leave the idea of the technological cognitive interest. There are moreover grounds for asking whether the specific combination of realistic and instrumental elements we came across above even needs a concept of a cognitive interest at all, whether the sound content of the theory is safely preserved in the theory of a primary level of instrumental access to nature. There is at any rate no attempt to stretch a theory of the cognitive interests beyond the approach I have outlined here. It is reasonable to believe that Habermas's critique of subject–object philosophy implied that at least his social philosophy could not be developed in close contact with a theory of the cognitive interests. Furthermore Habermas's subsequent important distinctions in the theory of science, such as the distinction between reflective and reconstructing types of science, bear no relevance on a theory of technology.

In order to trace the theme of technology in his later social theory we must therefore look at the critique of the subject–object philosophy, but I would first like to study his treatment of technology in this early phase in a more social perspective.

# 7 Technology as Ideology and a Form of Action

In retrospect, the theory of cognitive interests has come to comprise a relatively isolated and incomplete part of Habermas's authorship. However, the same can definitely not be said of his other writings on technology from the 1960's, which are the subject of this chapter. Here we come across intuitions and themes that are pursued in a different form in his theory of communicative action. The reason why these ideas needed reformulating and the way in which they are being developed today are, as I mentioned in the introduction to this part of my book, dealt with in chapters 9 and 10.

Chapter 7 falls into two sections. In the first I shall be looking at the most political themes, and concentrating in particular on Habermas's critique of the so-called technocracy thesis and his view of the ideological role that technology plays in modern society. The second part deals with his first attempt at a systematic theory of society, which was based on the distinction in the theory of action between labour and interaction.

I. The form of the technocracy thesis with which Habermas deals seems to be specifically a product of postwar Europe. It is highly distinct both in form and content from the theories put forward by the American technocracy movement from the interwar period, the best-known representative of which is Thorstein Veblen. This movement was for the most part interested in breaking down irrational obstacles to the growth of machine technology and the mass production schemes of large-scale industry, i.e. technology in the more traditional sense of the word. It concentrated therefore on the engineer in particular — an occupation that the movement took to be the driving force behind a rationalization of society — and was primarily an ideology of, for and about this particular occupational role. Its faith in progress and the future was optimistic, often verging on utopian. The movement can be regarded as a particularly narrow form of the tendency towards rationalization which is always present in modern cultures. (For the political version of the American technology movement, see Winner 1977, chap. 4.)

The European, postwar version in Germany was represented by Hans Freyer, and more importantly Helmut Schelsky (Schelsky 1961), both of whom were inspired, to a certain degree, by the writings of Jacques Ellul. This theory not only aims to pinpoint the characteristics of modern technology, but also maintains that it is modern technology that determines the basic character of modern society. In this variant, technological rationality, in the narrow sense, as rational interaction with the means of production, merges with Weber's theory of the means–end rationality of the bureaucracy. The catch phrase "objective exigencies" (*Sachzwang*) covers both fields. The analysis essentially differs from Weber's through its attributing less importance to the decisionism of the politicians. It is therefore not primarily interested in the engineer, but rather in the bureaucrat. These two figures then merge in the conception of the public expert within all fields of public bureaucracy and administration. After the Second World War it was certainly natural to place particular emphasis on the role of the military-technological bureaucracy in government offices and large corporations, and here the American experiences of course played a fundamental part.

Nevertheless, the theories of Freyer and Schelsky have a clearly German tone to them. Here we find not so much technological optimism and utopianism, as a conservative anxiety over the lack of soul and meaning in the modern world. This sort of thought was of course an essential element in the general and cultural world picture in the 1950's and 1960's. Regardless of how critical one might be to the actual thesis itself, it undeniably touched upon themes and developmental trends that greatly concerned people on both the left and the right wings. The fact that these two authors perceive this process as inevitable does not reduce the impression that we are merely being served a new version of the traditional German critique of culture.

Perhaps it would not be stretching things too far to view the technocracy thesis as a reaction to the voluntarist elements in German political philosophy. The pendulum has swung from the notion that politics should be rooted in will and action, to an abandonment of the existential moments in the name of objective exigencies. For a further discussion of this debate, see Glaser 1972.

Habermas, too, seems to perceive the technocracy thesis as a variation on conservative philosophy, according to a text from 1968. He distinguishes between three different interpretations of technological progress, *"Praktische Folgen des wissenschaftlich-technischen Fortschritts"*, in Habermas 1971b.

The *liberal* interpretation regards technology primarily as providing relief and safety; it helps human beings to free themselves from the hardships and burdens which they have suffered throughout history (p. 339). The *conservative* interpretation draws attention in particular to the way in which technology assumes the character of an autonomous system, which rules out the possibility and the ability of the individual to decide his or her own life (p. 341). The *technocratic* interpretation, Habermas's designation for Schelsky's version of the technocracy thesis, simply takes this conservative way of thinking one step further. It is not only technical-scientific progress that is ruled by an immanent dynamic, but, with time, also the state apparatus within which it develops. State bureaucracy assumes the form of technological control (p. 343).

Habermas's dismissal of this form of the technocracy thesis — and it is the only one he deals with — can basically be seen as a critique of ideology. It is thus very tempting to regard this (as Honneth points out in Honneth 1991, p. 249) as a parallel to his critique of positivism. Both critiques represent a critique of ideas that are *both* an expression of a growing alienation and reification of the world — as Weber, and before him, Marx, depicted it — *and* which themselves contribute to reinforcing this tendency. Habermas himself describes the thesis as a manifestation of a new level of rationalization, rationalization level "number 2" with reference to that described by Weber. The entire body of public administration is now united in the implementation of strategies that build on a new technology, resulting in a strictly hierarchical, bureaucratic form of state leadership, where division of labour rules, and which is staffed by professional lawyers, (*"Verwissenschaftliche Politik und Öffentliche Meinung"*, in Habermas 1971a). Honneth states:

> The structural differences that he [i. e. Habermas T. K.] has drawn conceptually between communicative and purposive-rational action he now repeats at the level of the social process of reproduction to the extent that he distinguishes between social spheres according to which of the two types of action predominates within them. He thus unintentionally lets the analytical distinction pass over into a difference between empirical domains of phenomena so that in the end the fiction is produced of a society divided into communicatively and purposive–rationally organized domains of action (Honneth 1991, p. 255 ff.).

I nevertheless believe that Honneth fails in his attempt to apply Habermas's words about the relationship of positivism to empirical research: "the false consciousness of a correctly perceived practice" to the technocracy thesis (Honneth 1991, p. 249).

The essence of Habermas's criticism of the technocracy thesis as an ideology is not the claim that this thesis embellishes and legitimates an alarming development, nor primarily that it helps to conceal it. It would be more correct to say that this thesis contains an erroneous claim about the actual development and that it, against the background of this misunderstood developmental tendency, then puts forward a mistaken interpretation of society as a whole. The technocracy thesis misinterprets *first* the new conditions for the accumulation of private capital as a technical and bureaucratic process, which follows an immanent technological logic, and *then* takes this process to be an irreversible development of objective exigencies in all fields of society.

The point of the technocracy thesis is that it is an impossible Utopia; it is not primarily an omen of and an, albeit unwilling, acceptance of a state we should be trying to avoid at all costs, but a claim that we are situated in, or on the verge of, a state that cannot become reality. On this point Habermas's dismissal is sweeping and quite condescending:

> But the weaknesses of this technocratic model are evident. On the one hand, it assumes an immanent necessity of technical progress, which owes its appearance of being an independent, self-regulating process only to the way in which social interests operate in it — namely through continuity with, unplanned, passively adaptive natural history. On the other hand, this model presupposes a continuum of rationality in the treatment of technical and practical problems, which cannot in fact exist (Habermas 1971, p. 64).

Whether we perceive the technocracy thesis as a threat or a promise, a Utopia or a despot, is of lesser importance, as it nevertheless predicts something that is anthropologically *impossible*, namely that humans can also be permanently relieved of *activity* on the basis of given preferences and values.

> For the scientific control of natural and social processes — in a word, technology — does not release men from action. Just as before, conflicts must be decided, interests realized, interpretations found — through both action and transaction structured by ordinary language. Today, however, these practical problems are themselves in large measure determined by the system of our technical achievements (p. 56).

We are not threatened by a merging of technical and political questions, but by the lack of awareness about the political priorities in technical decisions. This is the context in which the technocracy thesis functions ideologically. By suggesting that this merger has already, or has almost, taken place, it conceals the political and economic interests which are very much alive and kicking.

In a somewhat less schematic rejection, which associates the former East-European state socialism with the technocratic tendencies in the West, Habermas indicates how this kind of confusion may have arisen. He takes advantage of the distinction, which is the subject of the second part of this chapter, between technical subsystems and the institutional framework of society. Marx's thesis about the fettering of the relations of production through the productive forces is explained by Habermas as a thesis that the relations of production have only passively adapted themselves to our progressive and active domination of nature. Starting from the most variant analyses of the deficiencies in the social organization of production, and not only by placing all blame on the right to private ownership and the accumulation of capital under the direction of atomized producers as our point of departure, we may reach the conclusion that the institutional forms themselves must be regulated according to a pattern of technological rationality (Habermas 1971a, p. 352).

In so doing, however, we overlook the fact that the form of action and rationality, which lays the foundation for the institutional framework, is also the only one that can accommodate the social norms and therefore the possibility of a criticism of political as well as technological domination.

I do not find it reasonable to say that Habermas's view, as I have outlined it here, conceals the role technology plays in modern society, nor that he comes to a standstill at a purely means–end model of technology, which I think everyone today would agree is oversimplified and outdated. Habermas does not deny here that technological systems, both within and outside of the government, contribute to defining their own objectives and developing imperatives. The point is simply that when such goals and values are more or less explicitly formulated and implemented, then we are irreversibly back in the realm of action and politics.

Habermas's point is that the legitimation of these kinds of values and objectives does not and cannot belong in the realm of technological rationality. It belongs instead to a process of legitimation where inherited value judgements and the articulation of new needs must meet in what he called a pragmatic model for politics ("Verwissenschaftliche Politik und Öffentliche Meinung", in Habermas 1971, p. 66).

This does not have to be interpreted as an anti-technological attitude per se, as becomes apparent if we take a step beyond Habermas and note that the generation of new and legitimate needs on the basis of new expectations and opportunities, as well as on the basis of needs that have already been satisfied, is a fundamental anthropological constant. We could not realize the possibility of an improvement of the human lot, if a process like this were not seen as basically legitimate. The really progressive role that technology has played in modern times consists in having paved the way for these very structures in the human consciousness in a constant interaction between increased satisfaction of needs and increased needs. (For an account of this see Blumenberg 1973, chap. 2.)

Paradoxically enough it is the impossibility of realizing that tendency, which the technocracy theorists maintain is about to break out, which renders possible the ideological function of the technocracy thesis. This is the quintessence of Habermas's catch phrase that technology and science have become ideologies in modern society.

I am not asserting that this cybernetic dream [Habermas is referring to Hermann Kahn's future research in particular here, T.K.] of the instinct-like self-stabilization of societies is being fulfilled or that it is even realizable. I do think, however, that it follows through certain vague but basic assumptions of technocratic consciousness to their conclusion as a negative utopia and thus denotes an evolutionary trend that is taking shape under the slick domination of technology and science as ideology ("Technik und Wissenschaft als Ideologie", in Habermas 1971, p. 118.).

As an ideology on a new level of rationalization, the technocracy thesis does, however, distinguish itself from earlier forms of ideology in various respects. Its legitimating function does not imply that it leads us to expect or accept an illusion of justice in an unjust world. It does not even attempt to bridge the gap between ideals and reality, which would make it guilty of the usual accusations in the traditional strategy of the critique of ideologies, namely of claiming that the ideals of justice and equality in the classical legitimations of the existing society remain valid, even if unfulfilled. On the contrary, it leans towards eliminating *these* kinds of *abused* moral standards by referring to the inevitability of development and its functional effectivity (Habermas 1971, p. 112).

Time seems to have been as tough on the technocracy thesis as it has been with its critics. Habermas's slender hope of a debate between experts and lay-people on the application of technology within the confines of a re-politicized public sector was thwarted by an overpoliticization of science in the Marxism of the 1970's and in some of its repercussions. Moreover the ideas of the protagonists of the technocracy thesis relating to objective exigencies in the public administration were ridiculed with the advent of neo-liberalism in the 1980's. This ideology legitimated a quite new and almost social-Darwinistic "objective exigency" in the pure development of the market forces, where the good old needs of private accumulation replaced all the talk of the objective trends in and towards state control and planning.

Perhaps, then, the critics of the technology thesis had the last word, at least to a certain extent: Hiding behind objective exigencies were not objective processes, but specific constellations of interests.

There is nevertheless one specific element in Habermas's writings on technology from the 1960's which undoubtedly still deserves attention, not least because of the fact that the development of information technology has led to new forms of technological utopias, where immediate and profound cultural and anthropological changes will supposedly be triggered by individual technologies in this field. I am referring here of course to his conception of the way in which technologies and scientific theories are transposed to and influence the lifeworld.

Not least through the previously mentioned catch phrase "technology and science as an ideology", Habermas has contributed greatly towards identifying these concepts in the radical public opinion of the 1960's and 1970's. Nevertheless it is incorrect to say that Habermas put forward a theory of *the* "relationship" between technology and science. It is of course difficult to say whether it is even reasonable to talk about *the* relationship between technology and science at all, and, if it does exist, what a comprehensive theory about it would have to include: The epistemological relationship between two areas of knowledge? The relationship between formulations and applications of knowledge? The changing historical relationships of dependence and the direction of the flow of information between technique and science? The various social relations such as forms of education and types of institution within two different social subsystems? I, for one, am certainly highly sceptical towards the existence of a comprehensive theory of these two fields, and am fairly confident that if one does exist, it would hardly be of any value. There is certainly no one theory in either of these fields that has any real claim to the position of *the* theory of *the* relationship between science and technology.

Habermas has made two major contributions to this complex set of problems. I dealt with those concerned with the theory of science (in detail) in the previous chapter. The second area in which he made his mark is connected to the mediation of technical and scientific information, the content of natural-scientific theories, to society at large. Habermas used the expression "science and technology" as a general term in this debate only in one specific and very narrow context, which is purely sociological, and not in the least epistemological. In modern societies science and technology are fused together in the sense that, from the point of view of the lifeworld, the development and distribution of scientific and technological knowledge can be said to take place within a single social subsystem. This must be distinguished from his fairly obvious point that experimental physics is limited by the test conditions that the prevailing technology is able to provide.

Natural-scientific knowledge, according to Habermas, does not enter the lifeworld as *theory*, but only in an applied form thereof, through its exploitation for technological purposes. Scientific information will never become directly action-oriented, it is only when it is transformed to and incorporated in marketable objects and technological projects that it becomes a part of our lifeworld.[20] This is what makes it correct in one context to speak of technology and science as a collective term, which together constitute the back-

---

[20] We should perhaps, strictly speaking, say "seldom", rather than "never". Social-Darwinism is an expression of an ideology that legitimates itself directly on the basis of a misunderstood natural science.

ground for one form of ideology. The significance of science for the lifeworld is expressed primarily through its practical and economic application, not in the form of the information it provides about nature, i.e. the semantic content of its theories.

II. The catch phrase "technology and science as ideology" was without a doubt one of the two pairs of concepts that put Habermas on the map, not only in the world of academic philosophy, but also in the public, cultural and political debate towards the end of the 1960's. The other was the distinction between labour and interaction.

His distinction between labour and interaction formed the point of departure for the formation of a theory of social philosophy where Habermas at one fell swoop distinguished between *two* forms of action (communicative action and purposive-rational action) and *two* fields of society (the institutional frame and the subsystems of purposive-rational action), with the result that each form of action was attributed to its own field of society.

I shall start off (a) by attempting to demonstrate the way in which this distinction is decisive for Habermas's first positioning of technology in a context that is closer to social philosophy than to politics. I will also be looking at Thomas McCarthy's demonstration of terminological confusion in Habermas's writings. In connection with this critique I shall (b) link McCarthy's critique to Honneth's, and ask some critical questions about the approach to technology in which this theory appears to result. Finally (c), taking the work of the German sociologist, Hans Linde, as my starting point, I shall show that it is quite possible, indeed, even advantageous, to make way for a concept of object-mediated action, or, to put it another way, a concept of materiality, also outside the field of purposive-rational action.

I would like to underline for the sake of clarity that this topic is of more than merely historical interest. The distinction between labour and interaction can be seen as a first draft of the distinction between lifeworld and system, on which Habermas's theory from 1981 is based. It is therefore necessary to investigate whether there are assumptions built into this theory right from its very conception that act as an obstacle to a satisfactory treatment of technology. This would appear to be Honneth's thesis. Regardless of this, however, it should become apparent in my account that technology as a theme has always been present as an essential element in Habermas's theory of society.

(a) The distinction between labour and interaction was first developed in association with a reinterpretation of Hegel's early philosophy. Habermas believed he could prove that the young Hegel distinguished between several, mutually irreducible patterns in his account of the development of the spirit.

(Cf. *"Arbeit und Interaktion"* in Habermas 1974.) Habermas later emphasized once again the significance of Hegel's early philosophy for the history of philosophy; it constituted, he claimed much later, an overlooked way of avoiding the subject–object philosophy, which was to continue to determine both Hegel's thought and subsequent European philosophy (Habermas 1985a, p. 32 f., see also the next chapter.) Furthermore the distinction between labour and interaction is without a doubt a rough outline of the later theory of a differentiated concept of rationality, in the same way as the distinction between the institutional frame and the subsystems of purposive-rational action was the first expression of the intuition behind what was later to develop into the distinction between lifeworld and system.

A closer study of the philosophical roots of these distinctions is unnecessary in this context. My concern here is the theoretical construction into which this distinction was introduced. The decisive point in this construction is the sharp distinction between the two forms of action and the two spheres of society, and the equally rigid identification of each of the forms of action with one specific social sphere.

Habermas writes that the most *general* distinction is the distinction between the two social spheres:

> So I shall distinguish generally at the analytic level between (1) the *institutional framework* of a society or the sociocultural life-world and (2) the *subsystems of purposive-rational action* that are "embedded" [*eingebettet*] in it (Habermas 1971, p. 93).

This distinction itself however seems to be derived in turn from the corresponding distinction on the level of the theory of action:

> In terms of the two types of action we can distinguish between social systems according to whether purposive-rational action or interaction *predominates*. The institutional framework of a society *consists* of norms that guide symbolic interaction. But there are subsystems such as (to keep to Weber's examples) the economic system or the state apparatus, in which *primarily* sets of purposive-rational action are institutionalized (all italics are mine, T.K.) (p. 93).

As the key words make clear, Habermas seems to be paving the way for a relativization of the rigid one-to-one association between social sphere and form of action. On the general, analytical level it is nevertheless the identification of each of the social spheres through one type of action that holds the field. Habermas develops this dichotomy further by attempting to establish its existence on a number of other levels: Language types, rules of action, forms of rationalization, and even patterns of internalization (p. 93).

A scheme of this kind does of course at first glance arouse the suspicion that this construction must overreach itself by incorporating at the same time distinctions on the most abstract levels of theory as well as on the more concrete levels.

For the time being, however, we can content ourselves with registering that his social theory in this early form rests primarily on the fact that the distinction between labour and interaction is conceived as a theory of action. There are two sources of his inspiration for this theory. The most obvious influence is Weber. Habermas himself introduces his theory as a "reformulation" of Weber's rationalization thesis, a formulation which is to "go beyond" and be based on "categorial preconditions" other than those Weber used in his original version (p. 91). In keeping with Weber's original intentions, Habermas attempts to extend his approach to act both as a model for the emergence of modern society and as a diagnosis of contemporary life. The distinction between labour and interaction is introduced to illuminate and provide a sufficiently differentiated and nuanced picture of the complex process which Weber called rationalization (p. 94). The development towards a modern, and for Habermas therefore also market-capitalist society, is marked by the fact that the general pressure to adapt, which the productive forces *always* exert on the institutional framework, results in the subsystems of purposive-rational action also taking over the basic legitimating function in society (p. 96). In this day and age, this process is expressed in the tendency to conceal, but not negate, the dualism between labour and interaction, (p. 103), as we have already registered in connection with the technocracy thesis.

It is however perhaps more surprising that Habermas wishes to see himself as also going beyond Parsons in the same way. This is not only because he wishes to use Parsons's ideas as evidence to support the notion that all general theories within the tradition of academic sociology deal with the transition from pre-modern to modern societies. Habermas is also indebted to him in his interpretation of the interaction-governed social sphere, i.e. the institutional framework of society, as a norm-governed system of communicative action. This distinction is without a doubt based on the intuition that forms the basis for the later distinction between lifeworld and system, but in this case we are dealing with two types of system in confrontation (Honneth 1991, p. 251).[21]

---

[21] The frequent accusations against Habermas that he, from 1981 on, polluted his hermeneutic approach with a system-theoretical element are thus misfounded. Indeed in 1968 the hermeneutical world itself was defined as a system. Honneth may thus be correct in his claim that at that time it was of lesser importance that Habermas, as opposed to Parsons, did not believe that the system of norms in a modern society was built on a

The following passage provides the most exhaustive presentation of this basic theory of action:

> By "work" or *purposive-rational action* I understand either instrumental action or rational choice or their conjunction. Instrumental action is governed by *technical rules* based on empirical knowledge. In every case they imply conditional predictions about observable events, physical or social. These predictions can prove correct or incorrect. The conduct of rational choice is governed by *strategies* based on analytic knowledge. They imply deductions from preference rules (value systems) and decision procedures; these propositions are either correctly or incorrectly deduced. Purposive-rational action realizes defined goals under given conditions. But while instrumental action organizes means that are appropriate or inappropriate according to criteria of an effective control of reality, strategic action depends only on the correct evaluation of possible alternative choices, which results from calculation supplemented by values and maxims.
>
> By "interaction," on the other hand, I understand *communicative action*, symbolic interaction. It is governed by binding *consensual norms*, which define reciprocal expectations about behaviour and which must be understood and recognized by at least two acting subjects. [...] While the validity of technical rules and strategies depends on that of empirically true or analytically correct propositions, the validity of social norms is grounded only in the intersubjectivity of the mutual understanding of intentions and secured by the general recognition of obligations (Habermas 1971, p. 92).

I am choosing to overlook one potential objection to Habermas's distinction. Undoubtedly, no analysis of an actual labour process or a specific technology in use can avoid the communicative and norm-regulated connections between the workers (and probably also their superiors). Seen from this angle it is likely that these concepts are of no *immediate* use to historians of technology or sociologists of labour. The concepts would in any case have to be mediated via a general historical analysis of the relations of labour throughout the European process of modernization. Owing to the fact that Habermas is so obviously pursuing a different goal, however, I shall ascribe this objection little relevance in this context.

Thomas McCarthy, by contrast, provides a more profound critique (McCarthy 1978, p. 24), which touches upon two issues. First he points out

---

cultural inheritance, which was accepted as a matter of course and beyond any debate, but was itself also on a level of reflected conceptions of norms.

that it is unfortunate that Habermas seems to want to perceive instrumental action and rational choice, the latter being a term he uses as a synonym for strategic action, as *two* different *types* of action. This sheds a strange light over the very concept of purposive-rational action, and furthermore it is somewhat inadequate. The derivation of goals or partial goals in specific situations on the basis of superior preferences, and the discovery of possible adequate technical means to attain these goals, are of course two different processes, but, as McCarthy points out, they are not different *types* of action. It would be more fitting to refer to them as two moments in a single type of action.

The gravest weakness in this terminology, however, lies in the fact that it remains an open question whether purposive-rational action is also to include instrumental use of technical objects (what we can call labour) or not. The first interpretation would clearly be most in keeping with the distinction between labour and interaction. The concept of purposive-rational action, however, appears also to be capable of including purely social strategies, such as those we come across in economic theory. That is probably the reason why Habermas vacillates between the terms rational choice and strategic action. This brings us nicely to the crux of the matter: There is confusion as to what exactly communicative action is to be delimited *against* and what it itself constitutes a counterpart to.

It may help at this point to draw in Weber's distinctions which Habermas uses as his starting point. Weber distinguishes first between social and non-social action, defining social action as follows:

> Action is social in so far as by virtue of the subjective meaning attached to it by the acting individual (or individuals) it takes account of the behaviour of others and is thereby oriented in its course (Weber 1968, p. 88).

The reason why Habermas attaches such importance to the distinction between purposive-rational action, understood as *labour*, and interaction, is that it is his goal to arrive at this very distinction in Weber, although he here also paves the way for technologies for control of people, and not merely of things (McCarthy 1978, p. 28). The purpose of this distinction is obvious and is as apparent in his critique of Marx as it is in his critique of the conservative technocracy thesis. A concept of historical rationalization built on the rationalization within the purposive-rational sphere alone would be hopelessly inadequate. It would not contain any normative standards for a criticism of the course of history and would be likely to degenerate into a purely bureaucratic administration. This is explained in greater detail in the critique of subject–object philosophy in the next chapter.

The reason why Habermas distinguishes between strategic action and communicative action, by contrast, is that he is thus able to make a distinction within Weber's general concept of social action, for purposive-rational action is, in Weber's view, a form of social means–end action. Habermas's purpose with this distinction is quite different to the one I mentioned above. I am tempted to call it Kantian, as it rests on the distinction between treating others as an end, not only as a means. I find this tenable despite the fact that Habermas of course envisages communicative action as governed not only by norms accepted through rational reflection in Kant's sense, but also by traditionally sanctioned and hermeneutically received moral views.

This ethical theory is not the contested issue here. In this context it is more important to note that the two distinctions Habermas implicitly discloses are each perfectly plausible on their own. The construction as a whole, however, does not function as a theory of society due to the aforementioned identification of types of action with social fields. If we retain purposive-rational action as a general category, the theory of rationalization suffers, as McCarthy indicates. It is impossible, on the basis of the theory from 1968, to distinguish in a historical perspective between the rationalization within the sphere of labour, the development and institutionalization of the technological drive in capitalist societies, on the one hand, and the rationalization of those spheres of society, on the other, determined by traditional morals and religion, such as the rationalization of the legal system and the development of empirical social science. These are themes that are given a great deal of attention in Habermas's later *Theorie des Kommunikativen Handelns* (Habermas 1987b). Furthermore, and this is the weightiest objection, the field for strategic social action thus appears norm-free, which is of course ridiculous. Strategic action takes place within the framework of accepted norms that act as restrictions on freedom of action and as resources for implementing interests. If we have learned anything from investigations into bargaining, it is that strategies and norms are intricately interwoven (Elster 1989, chap. 2).

The notion of a norm-free social sphere (complementary to a norm-free technical field) is merely the abstract counterpart to an idea of a society held together by norms alone. But what is the significance of this for the question of technology? It is at this point that it might be useful to pad out McCarthy's observations with some of Honneth's ideas.

(b) Honneth does not aim to take part in the traditional debate about whether Habermas's concepts can be adapted to allow a more concrete analysis than the general analyses provided by the theory of action. Instead, he makes the point that Habermas himself undertakes an exaggerated concretization through connecting the two types of action with different spheres of society.

This is what causes the bifurcation discussed above. Society qua lifeworld becomes a form of purely norm-regulated sphere. Since the government and the economy disappear into the purposive-rational subsystems, it would appear that the concept of power is irrelevant for the former sphere. Moreover, the fact that the purposive-rational subsystems are located outside the norm-regulated field leads to the emergence of a "fictious" norm-free, purely technical sphere. In this way Honneth maintains that Habermas involuntarily has re-employed a basic assumption from the technocracy thesis that he was in fact aiming to disprove.

> The structural differences that he has drawn conceptually between communicative and purposive-rational action he now repeats at the level of the social process of reproduction to the extent that he distinguishes between social spheres according to which of the two types of action predominates within them. He thus unintentionally lets the analytical distinction pass over into a difference between empirical domains of phenomena so that in the end the fiction is produced of a society divided into communicatively and purposive-rationally organized domains of action (Honneth 1991, p. 255).

But just how convincing is Honneth's critique? It might be useful at this point to distinguish between different levels in Honneth's critique. His most general contention is that Habermas confuses types of action and social spheres. Indeed Habermas himself admitted later that this was untenable (Habermas 1991a, p. 254). On a more concrete level this leads to the absurd idea of a society or a lifeworld, as it would come to be called later, without any form of strategic action and exercising of power.

Whether the second part of Honneth's critique is equally convincing, however, is a different matter. Honneth does make one further point here. Habermas uses the distinction between labour and interaction to establish a theory of the processes of modernization in modern societies, something he at that time still regarded as a part of a project he called a history of the species, a complete history of the human race. According to Honneth, Habermas, on the basis of his fundamental dualism, must grasp:

> [...] the emergence of capitalism ... as a breakthrough of purposive-rational principles into a lifeworld previously organized communicatively (p. 262).

Furthermore, in Habermas:

Processes of social domination — indeed, the problem of the social formation of power in general — are secondary for the model of the history of the species that leads to this practical conclusion. [...] But power and domination that disruptively influence the species' process of will-formation results not from the administrative control of socially privileged groups but also from the pressure for adaptation that purposive-rational organizations socially exercise (p. 267).

What we are seeing here is, he claims, basically a conflict between two different types of action, two spheres of society, but not between social groups.

This might appear to represent a convincing interpretation of Habermas's general scheme of history at this point in time. We are left with the impression that we are witnessing a struggle between technological rationality and morals. I, for one, cannot however see that this necessarily implies that Habermas inadvertently defines technology as an independent and norm-free field, as happened in the technocracy thesis. First and foremost, neither Marcuse's idea of technology as ideology or the theory of early Frankfurt School about state interventionism, which Honneth perceives as Habermas's sources of inspiration, would lead to this conclusion. Marcuse's theory, at least, was in fact built on an overpoliticizing of technology, as we saw in the previous chapter. Furthermore, Habermas did not accept the development predicted by the technocracy thesis as real, but regarded this very prediction as the ideological manifestation of a different and extremely worrying development.

Finally, it is not necessary to interpret Habermas's concept of purposive-rational action as power-free *per se*, as Honneth claims. Included in this concept, as we have already seen, are both instrumental rationality and strategic action, which in fact presuppose given objectives to be fulfilled and values to be maximized. Habermas has never mentioned anything as absurd as strategies that are independent of power. Indeed the relationship between norms and power is rather diffuse in Honneth himself. No doubt it could seem as if there was no room for power in Habermas's concept of the lifeworld from that period, as this excludes strategic action. It does not however follow from this that the sphere for purposive-rational action in its entirety is power-free, simply because technology itself is defined in opposition to moral norms. Despite the fact that the sphere for instrumental action has become independent of norms that are based on communicative relationships, this does not mean that it is free of attempts to uphold and expand acquired positions of power. There is at any rate nothing in the fundamental construction that forces Habermas to assert this.

This will become clearer if we, in accordance with McCarthy's revision, conceive technical and purposive-rational action as different moments of one single form of action which rules the purposive-rational systems of action. This sphere then no longer appears as norm-free in its entirety. This further allows us to accept that the same elements of action theory will also exist in economic, state-bureaucratic and technological decisions.

This rescue action does of course presuppose that we accept both McCarthy's revision and what I referred to above as Honneth's most general objection. Honneth however, also claims that the unintentional association with the technocracy thesis also exists in Habermas's later social theories, which build on the theory of communicative rationality. He locates it in the distinction between social and structural reproduction. In order to assess the validity of this point of view we must first look at the grounds for the change that occurs between the theories of labour and interaction and the subsequent concept of communicative action. This is, as I have already mentioned, the subject of the next chapter.

(c) Habermas's general views on the action theory were (and perhaps still are) strongly influenced by Weber. Before we go any further, I would therefore like to demonstrate that we also may arrive at a concept of materiality, if we take this theory of social action as our starting point. Habermas did not follow this train of thought in his theory in the 1960's. I am basing my arguments here on a work by the German sociologist, Hans Linde (Linde 1972).

Linde's point of departure appears to have been a great irritation over the fact that the actors' relationships to things and to each other *via* things do not seem to have received noteworthy attention in modern action theory.

It is interesting even at this early stage to look at the difference between the approaches adopted by Dag Østerberg and Linde. Both seem to have been annoyed by sociology's lack of interest in the category of *things*. Whereas Østerberg wanted to introduce his concept of materiality, which was inspired by phenomenology, as an opposition to Weber, Linde arrives at similar opinions by means of an immanent criticism of Weber. Østerberg therefore operates (or operated) with a distinction between social and material structures (Østerberg 1974, p. 26). Linde aims to accommodate the things in social relations and he claims primarily that they contribute to creating order and permanence in these relations. Although Østerberg's approach allows him to make striking distinctions, such as that between social structure as the ritual and material structure as the temple, the tendency ought perhaps to move towards apprehending the material as an element *in* social relations, for what is the ritual without the temple and vice versa?

I am however also critical towards Linde's conceptual apparatus. The clear-cut distinction between social relations on the one hand, and social structures or arrangements *(Verhältnisse)* on the other, seems methodologically unclear and unnecessary for the integration of things into action theory. It is not inconceivable that two actors who are trying to regulate their reciprocal behavioural expectations will have to do so through and attached to a technical medium, such as, for example, one that regulates the amount and rate of the flow of information. We do not need to introduce unclear, superindividual entities in order to understand this. As Linde himself says, Weber's category non-social action must not be taken as a form of action that excludes communication with people, but nor does it have to imply superindividual action.

In the following I shall be ignoring holistic elements of this kind in Linde's work, which appear to be the result of his confusing the need for a sociological treatment of the concept of things with Durkheim's thesis that social relations must be treated as things.

Linde starts off by introducing a double thesis about what he calls *Sachen* (as opposed to natural objects) defined as "all objects which are the product of human intention and labour" (p. 11). His thesis is as follows:

> The relationship towards objects are basic elements in the structure of human society, and for this reason ... should also be a basic category in the sociological analysis of these societies (p. 83).[22]

Linde is clearly not interested in the role of technology in the exchange with nature, but in its position in social relations.

The essence of the argumentation for this thesis is a relatively revisionist reading of Weber, in which he aims to demonstrate that it is possible to accommodate the idea of the reified character of social relations, or rather their character of being mediated by things, into Weber's project of a comprehending (*"verstehende"*) sociology of the meaning (*"Sinn"*) of actions and therefore also into his typology of types of action.

Linde takes Weber's distinction between social and non-social action as his starting point. Weber defined social action as action "oriented to...the behaviour of others", (Weber 1968, p. 112). In addition to this, Weber introduces two sub-groups of non-social action. Linde comments that since only the latter of these consists of phenomena relevant to society that are not available for a comprehending analysis (such as purely ecological or individual physiological natural processes), the *first* class of non-social action must be linked to specifically human activity. It includes actions in which people relate, even if only causally, to the actions of others:

---

[22] All translations from Linde are my own.

> "Without meaning" [in the Weberian sense T. K.] is thus in no way identical with "without life" or "not human" (Linde 1972, p. 41).

Within this class of non-social types of action, Linde is interested in particular in Weber's first type of examples — they include relations between actors that are causally mediated by, for example, machines. These kinds of artefacts are part of a social context that is comprehensible, i.e. available for an analysis based on *social meaning* as a basic concept, despite the fact that we are not dealing with social action here. This is because these particular artefacts are consciously produced and/or are included in the actions of the individuals as an end or means.

By distinguishing in principle in this way between the causal role of consciously produced products and that of purely natural objects or processes of the kind I mentioned earlier, Linde creates a *four-fold* typology, which is supposedly better suited to preserving the significance of the things and technology with respect to action theory, than Weber's. In addition to Weber's category of social action, we also have three types of non-social action. The second class of these is the most important; the two others are rather strange and are somewhat reminiscent of residual categories. We do not need to look at these two classes here. Class number two includes actions:

> ...where either a). the orientation of the actions originates in the action of others in so far as this orientation is determined causally but not through the meaning of these actions or b). the orientation is towards the artefacts or objects that are comprehensible as means or purposes of one's own actions and/or the actions of others (p. 42).

Linde also makes the interesting observation that Weber's thesis on categories from 1913 (Weber 1985) shows a greater willingness to treat relations to things, or, to be more precise, social relations that also include use of artefacts, within the framework of a comprehending sociology than the unshakeable distinction between social and non-social action in *Wirtschaft und Gesellschaft*.

In this thesis Weber describes how members of a rationalized society relate to that form of calculation and that technology which rendered such a society possible in the first place. This observation appears in the last part, which looks at how it is possible for forms of social relations to exist that are such that:

> the individual gets involved in social cooperation without any conscious participation (Weber 1985, p. 465).

In connection with the passage quoted by Linde, Weber develops a distinction between understanding and taken-for-grantedness: Weber refers to (p. 471) examples such as using a shot-gun, taking or even driving a tram. Even if scientific principles are involved in all these operations, no actual or conscious application of them is involved. The agents are only oriented towards expectations of coordinated behaviour which are linked to the use of these artefacts.

This train of thought invites several types of commentary. It is worth noting that Weber here touches upon the very same thoughts which we found in Gehlen's theory (Part I, chap. 2) that the primary manner of operation of technologies is their relieving function. The introduction of new technologies can at first come as a shock and have an alienating effect. But they soon become routinized, a process which has been called a "secondary traditionalization" (Hörning 1988, p. 84), in the general public who use the technologies, but who have no knowledge of the routines associated with the production and professional running of these technologies. This of course, on further reflection, paves the way for a feeling of alienation, albeit on a new level.

Linde's perspective, which I am following here, differs somewhat however. He reads into Weber's observation a recognition of:

> a fact which is decisive for our considerations, that objects in general have a potentially institutional function (Linde 1972, p. 49).

If this is the case then we are forced to ask ourselves: Which types of actions are of relevance in relation to things considered as instituitons? The significant point here is that Weber himself appears to shatter the conception of technology as pure means in a means–end rationality, at least to a certain extent. Since technologies are placed on a par with institutions, we must also realize that they themselves are capable of initiating new *goals*: Technology is no longer only *use of* techniques.

Linde's main point is thus that things primarily play their social roles as arrangements or regulations that not only define the subjective sphere of action between two actors, but also the positions it is possible for them to occupy on the basis of a social structure. They do not concern:

> ...ego and alter in their unmistakable subjectivity, but concern[ing] objectively norm-regulated positions, roles and status they have achieved or been ascribed (p. 54).

The very fact that social behaviour demonstrates such a surprising regularity, continuity and similarity, Linde claims with Weber, cannot be ascribed to the moments that remain subjective in the sense of constituting purely individual elements, i.e. the direct relations between two actors' intended or perceived meaning. They must either be a result of common interests or exist in "the case itself". It was on precisely this level, as we have already seen, that Linde located the sociological function of things. Once we have positioned things on this level, making them relevant for a description of the positions and relations that the actors can enter and assume, it then becomes possible to regard things and technologies as regulating in relation to all the types of action with which Weber operates. Through his ability to thus ascribe to things a "the social quality of both regulating and determining behaviour" (p. 59), Linde is also able to summarize more specifically what things actually determine:

> ...a) social positions including the element of behaviour and rank, b) patterns of behaviour that are neutral and diffuse as regards positions and c) finally ideas and expectations (p. 60).

The merit in Linde's work lies in his having studied the relationship between technology and norms of actions, and also indirectly in having shown that technology cannot be accommodated in a simple means–end model. I therefore believe that the issue of the relationship between technology and action theory can most fruitfully be developed further in two different directions.

With regard to the concept of things or the concept of materiality in general it appears that they can enter into types of action of any kind, regardless of whether we perceive things as bearers of a recognized order in systems that build on one particular type of action, or as a medium that connects two actors.

The other direction means asking the question whether there are specific types of action linked to modern technology, with its holistic character of networks, concatenations and hierarchies. Linde points out that technology is not only the means, it also creates new ends. It seems to have been generally accepted for some time now that the means–end model is too narrow. The earliest registration of this fact that I have come across was made by Thomas Carlyle in 1829, (cited in Borgmann 1984, p. 59). This kind of opinion has now become common-place. No doubt Lukacs deserves much of the blame for the fact that so much philosophy of technology within West-European Marxism has apprehended technology as the ideal type of the purposive rationality. This occurred when he identified Marx's concept of reification with Weber's action-theoretic analysis of rationalization, and thus planted this distorted form of rationality firmly *within* the sphere of production.

I will be returning to the question of Habermas's views on technology and types of action within the framework of communicative theory. Finally there are perhaps grounds for asking ourselves why Linde has been overlooked by so many theoreticians within this field. One argument might be that in pointing out that actions and interhuman relations are mediated via various physical media, he was merely stating the obvious. The only response to this is that it is surely better to put the obvious on the agenda than to continue to overlook it.

# 8 The Critique of the Philosophy of Consciousness

The *basic intuition* underlying Habermas's critique of the presuppositions of the philosophy of consciousness has in fact been present throughout his philosophy. As early as in the various essays published as *Technologie und Wissenschaft als Ideologie* in 1965, he distinguished systematically between labour, which was perceived as a form of instrumental purposive-rational action,[23] and interaction, which was linked to intersubjective and communicative relations mediated by language and concentrated around the reception and continuation of, for example, traditional elements of culture. This perspective was elaborated to form a critique of Marxism in the same works. A distinction of this kind is also presupposed in his early, and later quietly dropped, doctrine of the various cognitive interests and the types of science to which they correspond, in the form of a distinction between the technical and the practical interest (see also "Erkenntnis und Interesse" in Habermas 1987a, p. 301). It would thus seem reasonable to conclude that the idea of the necessity of a cognitive differentiation in the form of a theory of various types of rationality has always been present in Habermas's writings, and that it gained significance with time. It is however also true that these first attempts were incapable of being satisfactorily implemented until a different light was cast over them via the turn that Habermas's philosophy took in the early 1970's.

In his first inaugural address as a professor in Frankfurt in 1965 ("Erkenntnis und Interesse" in Habermas 1987a), Habermas declared that he intended to return to a theme that had its origins in Horkheimer's works from the 1930's and German idealism. This theme was the false impression of contemplative and disinterested reflection that he claimed to discover in the classical conception of theory in Western philosophy. As I have already mentioned, the main purpose of the theory of cognitive interests was to develop an anti-objectivist position. When he was reinstated as a professor in Frankfurt in 1984, he made it clear he no longer regarded it possible to continue to develop the tradition from the critical theory of the interwar period, at least in the form of a direct elaboration of the existing theory programme. "It is not my aim to continue the tradition of a school" (Habermas 1985b, p. 209). The possible theoretical presuppositions for a critical theory of society

---

[23] For a discussion of the confusion in Habermas's use of these concepts, see the previous chapter.

had changed to such an extent since the 1960's that this was no longer feasible. This brings us to the afore-mentioned break in Habermas's philosophy which also comprises the precondition for his critique of subject–object philosophy. (For Habermas's vocabulary in this respect see the introduction to Part III.) This revolution in his philosophy is often described as the linguistic turn.

For our purposes it might be most expedient to describe this turn in two steps, and see it first (I) as a self-criticism of his own early theory and of the critical theory in the early Frankfurt School and other forms of Marxism. This criticism can serve as a point of departure because it contains certain basic presuppositions, which all the theories I mentioned in the Part II share with their inheritance from Descartes and Kant. With regard to materialistic phenomenology this inheritance is of course primarily mediated via Husserl. By looking at this critique we will arrive at another level where we can then (II) identify the common elements of a subject–object theory in Marxism and in classical epistemology.

This turn, or break, is of paramount importance because it also implies a fundamental criticism of Habermas's own early philosophy. An investigation of the background to this turn in his thinking therefore comprises the best point of departure for an investigation of the subsequent critique of the philosophy of consciousness. It goes without saying that this turn has also a positive side in the application of modern pragmatic theories of speech acts (Austin, Seale) and social psychology (Mead), where it has contributed to the development of a differentiated theory of rationality. The expression linguistic turn can thus profitably also be narrowed down in the direction of a pragmatic turn. Within the scope of this book, however, I cannot give an introduction to the concept of communicative action, with which I am henceforth assuming the reader is familiar. (For a general introduction to the concept see McCarthy 1978, White 1988, and Cooke 1994.) I thus present this turn on the basis of the overall perspective in this book, namely theories of technology in social philosophy. My aim here is only to indicate how this critique undermines the theoretical presuppositions in early theories, such as that put forward by Sartre, and thus renders Habermas's own theory a potential foundation for a superior theory of technology.

(I) In connection with this turn we come across a certain ambivalence in Habermas's thinking. Is his critique of the theory of consciousness a systematic philosophical criticism, or is it a merely a criticism of it to the degree that we are interested in constructing a meta-theoretical basis for social science? He frequently states that the theory of communicative action is the model "best suited" or "worth using" in connection with a social theory. If we opt for the latter interpretation it is tempting to ascribe to him an almost instrumental model, albeit promoted to the level of meta-theory, where each individual science selects the models and tools that serve it best. This is, however, a very un-Habermasian interpretation; it is more reasonable to say that the choice of point of departure rests on a philosophical critique of the philosophy of consciousness, of the absurdities and inadequacies that are revealed if we employ it as our point of departure for a social philosophy. Nevertheless Habermas undeniably owes us an account of another point, namely whether there is any such thing as epistemology or a philosophical theory of the mind, whether there is, for example, any philosophically interesting theme that could be called the philosophy of perception?

A reasonable interpretation would be that the pragmatic turn does not imply a total relinquishment of the intentions that underlie the programmatic element in the original critical theory of Adorno and Horkheimer. Epistemology is only possible as a theory of society. The change that has taken place, however, is so fundamental that it is no longer a form of epistemology, but rather a theory of rationality founded on a philosophy of language, which constitutes the social-philosophical paradigm. The change in basic philosophical attitude which I have discussed is in fact a revolt against the general framework of classical epistemology, and not an adaptation of it to a social philosophy. It is reasonable today to attribute to Habermas the conception that no epistemology can become a social philosophy. The earlier idea that an epistemology was only possible as a social philosophy rendered itself impossible on the grounds of the theoretical assumptions that the final social philosophy *itself* adopted from the critique of epistemology. This kind of union of philosophy and social theory is only possible if we look for the junction in the immanent rationality in communicative interactions. There is only room for an independent philosophy here in the form of a separate function *within* the scientific division of labour, where philosophy functions primarily as an intermediary element. To the extent that there is a direct transfer of elements from the original thesis in critical theory, this must be seen at the outset as a theory of the role and function of philosophy in modern societies. "There is no longer any separate field for philosophy", it has become an interpreter and a direction indicator in debates about various meta-theories within and in the relationship between the various individual sciences. This is the message of an essay with the this title (Habermas 1983b). Philosophy is included in a cooperative with the other social sciences, as Horkheimer had envisaged it (Horkheimer 1931).

What we are now concerned with is replacing epistemology with a theory of the potentiality of rationality in communicative contexts, and developing and illustrating this potential by means of the theory of language in association with a theory of one particular type of society, namely the modern or post-traditional societies in which we live. Here Habermas is faced with yet another classic problem, with regard to connecting a philosophical and an empirical theory: How to connect the basic concepts in a theory of speech acts, which operate on a general philosophical level, to the concepts in a theory of modernity, which apply only to a specific epoch. It seems here that he attempts to find support in a theory at which Marx at least hinted, that the most general and abstract social philosophical categories become concrete and "practically true" through the historical development (Habermas 1987b, p. 402 and Marx 1953, p. 24 ff.)

These kinds of considerations are so fundamental for Habermas that he states in several places that he gave up his original project of completing a critique of an alienated world and social science *in the form of* a critique of method.

> Anyhow I did for a long period believe that a project of a critical social science had to vindicate itself in a methodological and epistemological way... I no longer hold this view, because the attempt to introduce the theory of communicative action through a methodological perspective, brought me to a dead end... The theory of communicative action, which I have since developed, is no continuation of methodology with other means... It has broken with the primacy of epistemology and treats the conditions for action coordinated by understanding independent of the conditions of the objectivity of experience (Habermas 1982b, p. 10).

An attempt at developing a theory of communicative action within the framework of a doctrine of method would mean remaining within the horizon of traditional epistemology. This is the reason why Habermas's own theory of cognitive interests and the quasi-transcendental philosophy of science from the 1960's that was linked to it had to appear obsolete to him. Although we can continue to employ a concept of positivism to describe a science that reflects and reproduces an objectivist illusion and an alienated reality, a "positivist dispute" hereafter can no longer be seen as a debate about methods. Besides, this was undeniably a conflict that had developed in a fruitless direction. Nor can his basic intuition, which underlies the distinction between labour and interaction, be expressed adequately within this frame. With its reformulation within the linguistic theory that Habermas developed in the early 1970's, it gained, as we know, a completely new form in the theory of a differentiation of various types of speech acts and forms of rationality.

What then are the main complaints against the basic philosophical view which Habermas sometimes calls philosophy of consciousness, and sometimes subject–object philosophy, and, in connection with Marx, also the paradigm of production? At the outset we are confronted with a sole subject, which relates theoretically, i.e. cognitively, to a world of things. In directions pursued by later social philosophy this model was then transferred to two levels. On both levels the idea of generative or productive processes will therefore assume a central position. The tradition of theories of consciousness, at least after Kant, i.e. in the part which bears relevance for social philosophy, is therefore a *theory of constitution*. First of all society is perceived as a collective subject or an aggregated result of individual subjects, which now relate to nature in a practical, processing way. This is Marx's concept of metabolism (*Stoffwechsel*).

This concept of a subject which relates to its physical surroundings in a primarily transforming way is obviously too narrow to serve as a basic model for understanding the entire field of forms of human behaviour. In addition, on the second level the relationship between the individual and society is later also perceived according to the subject–object model, in that society is seen as constituted through the output of the individual subjects. This brings us to another point which is most unclear: What is actually subject and what is object within this configuration, the individuals or the society? Society in this way is both the *subject* in relation to *nature*, and the *subject* in relation to the individuals that it constitutes and then the *object* for them, their own product. This leads to an unacceptable confusion with regard to which human relations and types of action should be emphasized in the various contexts. (I will return to the subject of confusion in the relationship between the individual and society below.)

These forms can be found in particular in the critique of the paradigm of production, in, for example, a discussion of Göry Marcus (Habermas 1985a, p. 95 ff.). In particular the first of the two elements I mentioned above is worked out in detail here: The paradigm of production treats all forms of human activities alike and reduces them to labour, and in so doing leaves no room for the establishment of explicitly normative aspects of and within the theory of society. We can thus develop a philosophy of history about a necessary succession of historical forms, but the paradigm does not provide us with a yardstick for an ethical assessment of them. We lack the differentation in types of rationality that is necessary in order to find the correct place for a form of moral discourse.

The association with the subject/object model thus also has consequences for the critical and political aspects of Marxism. This critique is formulated in Habermas's settling of accounts with Horkheimer and Adorno, and therefore also Lukacs (Habermas 1987b, vol. i, chap. 4). His main point here is that a theory that builds on the concept of the reification of the consciousness in the modern world is tacitly founded on a concept of consciousness borrowed from the philosophy of consciousness, and within this framework the intended critique of society cannot be expressed consistently.

The critique of this kind of subjective reason is also formulated by Adorno and Horkheimer, as a critique of the encroachment on nature which a subject like this must undertake in order to control it and secure its own self-preservation. Since they do not have access to any other alternative concept of rationality than that which they have already condemned as instrumental and alienated, they become caught in an apori.

(II) If we search back to the common features of the Marxist paradigm of production and the classical, epistemological subject–object model, we find two fundamental weaknesses. They appear most clearly when this tradition is used as the basis for a philosophy of society, as is the case not only in subsequent developments of Marxism, but also in phenomenologically inspired sociology, i.e. in the continuation of the tradition from Husserl, via Alfred Schütz, Peter Berger and Thomas Luckmann.

1). No sociologist worth his or her salt can possibly conceive of society as exclusively constituted through the subjects; we must also conceive of the subjects as being constituted through society to the same degree. This is a result of the fact that the original theory from transcendental philosophy is to be made operational for social philosophy, which entails an unavoidable step down, as it were, from the level of transcendental to empirical subjects. The point here is that we are now in a deadlock, since the subjects as well as the society are both constituted and constituting. We are obliged either to locate ourselves in the middle of a process that can never be fixed nor grasped, or to objectivate, pretty much at random, one of the extremes. This is the cause of the irritating oscillation that we observed in connection with Sartre's account of the relationship between man and machinery, and which can also be found in the works of Merleau-Ponty: A constant oscillation between the subjective and the objective factors, where the latter are only produced by the former, which in turn are constituted through the latter, and so on and so forth, ad infinitum. Nor does it help if we define society as a collective subject, or, in accordance with the requirements of methodological individualism, as an aggregate of the performances of the individual, empirical subjects. The dispute between individualistic points of view and holistic interpretations clearly bears no direct relevance in the perspective outlined here: It affects both camps. What is really strange in this undeniably weighty criticism is that Habermas here wins an unexpected ally in Foucault. This critique was anticipated by him in his dissection of the humanities in the last century in *Les Mots et les Choses*. (See Habermas 1985a, chap. IX).

2). There is more to this criticism than the mere fact that the relationship between the individual subject and society becomes unfixed: Within this framework it is impossible to explain the emergence of intersubjectivity. It would go beyond the scope of this book to give an account of Habermas's critique of Husserl's theory of the individual subject's constitution of an intersubjective environment (Habermas 1970/71, 2nd Lecture). The final result of this criticism is that we must abandon the constitution theory of society. The subjects do not relate to their surroundings, their traditions and contexts of meaning (and here we can interject: Their technological systems) according to the pattern of a subject that confronts or constitutes a foreign object. The fact that society from the outset requires shared opinions cannot be explained satisfactorily by this model.

Advocates of strict methodological individualism will perhaps dislike this formulation, which is intended methodologically and not as a cultural theory. As I understand Habermas, whose relationship to this debate is fairly chivalrous, it can clearly be replaced by the inclusion of the relations between people at this primary level in social theory (see also chap. 4). But this is not possible if alter is understood as a monologically constituted subject, even if alter is constituted as constituting. This is after all the consequence of this content, if not the method of Hegel's analysis of the "Kampf um Anerkennung"; the genesis of society must already presuppose a plurality of individuals.

We can then conclude that neither the relationship of subject to subject, nor the relationship level of the individual subject to the collective level and vice versa, gains anything by being apprehended as a *relationship of constitution*.

It is less significant that Habermas, in his criticism of Adorno's concept of mimesis, clearly overlooks the fact that the element of *aesthetic* rationality, which is also included in this concept, cannot easily be adapted to fit into his own general theory of communicative rationality. More important is his insistence on the paralyzing effect of a paradigm from the theory of consciousness on the critical, liberating and conciliatory aspects of the original critical theory in the beginning. Behind the change of theoretical paradigm, behind the transition to a linguistic point of departure, there therefore lies, in the final instance, a practical intention, and I find support for this interpretation in the following passage.

> My intention is to renew a critical social theory that secures its normative foundations by taking in the experience of thought gained along the way from Kant through Hegel to Marx, and from Marx through Peirce and Dilthey to Max Weber and George Herbert Mead, and by working them up into a *theory of rationality* (Habermas 1982a, p. 232).

Finally, in order to put the whole turn in its proper perspective, I am going to refer at first to an early article by Albrecht Wellmer (Wellmer 1977), and then supplement his account by a brief look at some of Habermas's more recent works.

Wellmer locates the break in Habermas's development in relation to the development of and problems within the Marxist tradition in general, from Marx, via Lukacs, to Adorno and Horkheimer. Marx's original philosophy of labour was part of a critique of idealism. His aim was to underline the importance of the active and changing nature of man's role in relation to his surroundings. According to Wellmer, however, this philosophy concealed a reductionist element in Marx's theory of practice. The point Wellmer is making is that the alleged dialectic between the forces of production and the relations of production, and thus also the very theory of the relations of capital as a social relation between people, was supposed to indicate a non-reductionist relationship between labour and social arrangements. A naturalistic theory is in itself insufficient for illustrating the political and normative aspects of the critique of the political economy and the theory of the revolutionary class consciousness. If Marxism was to continue to have any value for a diagnosis of contemporary life, then it would have to either prove that capitalist society was systematically undemocratic, or perceive socialism in purely technological, or perhaps even technocratic, terms — which of course was never Marx's intention.

The novelty in Lukacs's theory was his linking of Marx's economic analysis of the exchange of values on the market and the wage relationship to Weber's theory of purposive-rationality. This resulted in the labelling of Weber as an ideologist, since he had only stated and not criticized this fact. In keeping with the later Frankfurt School, Wellmer perceived Lukacs's critique of ideology as equally inadequate as the position it criticized, since it overlooks the fact that science and technology have become ideology, that capitalist society has moved beyond the sphere of the earlier liberal legitimating of capitalism on the basis of the theory of equal exchange on the market. The simple identification of this new form of ideology with purposive-rationality does not grasp the new role politics gains in relation to the basis.

The emphasis on the independent role of politics was central in the Frankfurt School in the 1930's. The emphasis on the fact that reification and ideology are more than mere crisis phenomena, however, resulted in this case in the interpretation of post-liberal capitalism as a qualitatively new and crisis-resistant economic system. In this way the critique was transferred to a general, superhistorical level, and the connection with contemporary analysis and critique was finally broken. Wellmer's main point, which was clearly intended as a defence of Habermas against attack from the traditionally Marxist camp, is that this connection can only be re-established by making the very change on the general philosophical level that Habermas makes. This is the driving force behind the original introduction of the distinction between labour and interaction. The rationality in the political and social fields is liberation and dissolution of domination, and linking this with a naturalistic concept of practice is itself ideology. This differentiation itself, however, necessarily takes us beyond the quasi-transcendental paradigm, which extended Kant's theory of constitution via a theory of labour. Playing communicative co-action and rationality off against instrumental rationality presupposes that the philosophy of language is made fundamental, and, in a theory of intersubjectivity, a constitution theory of society, which builds on epistemological presuppositions, does not occupy the position it does in the epistemology of, for example, Kant and Husserl.

I find further confirmation of this interpretation of the significance and function of the linguistic turn in Habermas's philosophy in his most recent book. While his prime goal in *Theorie des Kommunikativen Handelns* was to develop a theory, founded on communication theory, of differentiated forms of rationality, which would serve a descriptive theory of society, he is now striving to reformulate also the concept of *practical reason* in the direction of a theory of communicative reason (Habermas 1992, p. 17 ff.), that will serve as the basis for a philosophy of law and a theory of democracy.

Finally, then, I believe that there are grounds for emphasizing a more comprehensive form of *continuity* with the original critical theory, which does after all exist. Wellmer maintains that it is to be found first and foremost in the fact that this new approach paves the way for a new form of critique of idealism, in this particular case of a linguistic idealism. This kind of critique can affect both Gadamer's version of hermeneutics, and Peter Winch's version of Wittgenstein's theories of the relationship between language games and reality. The term "linguistic idealism" can be said to describe theories that allow linguistic communication to dissolve in self-understanding and which interpret all such understanding as unfettered. The anti-idealistic motives that are preserved from Marx's original materialism thus serve to underline the fact that the human subject is never completely accessible to itself, and that its attempts to achieve comprehension are always subject to external interference by social factors. In brief, we can say that materialism in this sense refers to the contingent initial conditions of all understanding. What still remains of Marx's concept of materialism is a problematization of the possibilities of communication in relation to a rationalized lifeworld and the differentiation of systems from the lifeworld. It is in this context that Habermas's communicative theory proves its worth as a means for putting technology under the microscope. What is lost in relation to the earlier theories we studied in Part II is that the concept of production no longer functions as a foundation and model for theories of interpersonal relations and rationality in general.

# 9 Technology between the Lifeworld and the Systems

The objective of the two final chapters is to investigate in detail how technology can be approached within the theory of communicative action and which unexploited resources for a theory of technology, if any, exist within this theory. First of all, however, let us summarize the most important results from the critique of the subject–object theory in the preceding, which may be useful for a theory of technology. As we saw, this critique led to the conclusion that the very idea of a constitution of society, based on the model of the constitution of the epistemological object, had to be abandoned. Communicative connections do not constitute society; they do not simply take over the *position* that was previously occupied by the performances of consciousness in an otherwise unchanged scheme. It is of course reasonable to say that within the concept of the constitution of society, or its self-constitution, in the works of Hegel and Marx, labour took over the role that self-consciousness had played in Kant's scheme. In the dialectical tradition the structure of labour and reflection united objectivation, alienation and the negation of the alienation in one constellation. We must therefore conclude that labour and technology, taken as an expression of instrumental action, can no longer be regarded as the pattern of rationalization of history or as the basic principle of reflection.

It is important to emphasize that it is this connection of a theory of rationality and historical rationalization with a theory of the structure of labour which is lost. The social problems and source of social criticism inherent in the critique of alienated labour and politically uncontrolled technology in modern societies will, by contrast, continue to be fully relevant, also within the perspective of a theory of communication. The critique of the subject–object theory is completely independent of whether one chooses to regard labour as being liberated or exclusively locked in the alienated, functionalized version that it exhibits, according to the Marxist diagnosis, in capitalist societies. Let me underline once more that my overriding aim here is to discover the extent to which a concept of materiality can be successfully formulated also within a communicative theory.

The transition in social theory from a subject–object model to a communicative model results in our having to ask ourselves the important question: On which level in the new theory and on according to which of its own basic concepts can the function and position of technology in a modern society can be determined. Now the theories of subject–object philosophies do give us as an indication in this respect. In the dialectical tradition of Marx and Sartre, and in a different version also in Gehlen's more superhistorical, anthropological approach, the position of technology in human society is defined as an element in the *reproduction* of society. The position of technology as a social phenomenon is located in the area where society as a collective subject reproduces itself through its metabolism with nature. Social structures and institutions are in their turn rooted in this relationship between society and nature. A theory of technology filling this position can be formulated on various levels: Both as 1) a diachronic theory of a) the role of labour in the origin of human society, and b) the role of technological change in the transition between decisive periods in these societies, and especially in modern societies, and 2) as a synchronic theory of a) a universal logic of development for these societies and b) with a view to understanding the reproduction of modern societies. Gehlen was, as we have already seen, primarily interested in presenting a theory of the problems within areas 1a and 2a; Marx dealt in part with questions pertaining to 1b, but concentrated mostly on those in 2b; Sartre looked almost exclusively at problems in the area 2b.

How then does Habermas's own conceptual grid relate to these problems when they are put in this way? The actual expression "reproduction of society", which undeniably bears a striking resemblance to some of the philosophical presuppositions of the subject–object theory, is not in fact a key concept in *Theorie des kommunikativen Handelns*. Like all other general social theories, however, Habermas's theory of communicative action must also be capable of answering the question of how a social structure is even possible at all. One may say, then, that the set of problems dealt with in the traditional theories — in association with the concept of reproduction — in Habermas appears in connection with and within *the* relationship *between* the two forms of integration with which his theory operates, namely *social* integration and *system* integration. His treatment of these two forms of integration will therefore be the main topic of this chapter. The concepts of these two forms of integration were not, however, introduced directly by Habermas. They can only be arrived at via the two concepts whose reciprocal relationship must be said to constitute the cornerstone in that which, in a narrower sense, constitutes the social-philosophical part of Habermas's theory: The idea of the distinction between the concepts of lifeworld and system.[24] This is the conceptual frame in which the earlier

---

[24] I am not trying to make unnecessary distinctions in a multifarious work here. I do however feel I am justified in distinguishing between the theory of the lifeworld and system, which in a narrow sense can be called a social theory, and which is dealt with

problems relating to labour, rationalization and alienation can be discussed on the basis of new premises.

The background for the progression in this chapter is as follows: I shall start off (A) by briefly considering what a communicative theory is able to tell us about the role of technology in association with the emergence of human society (what I called area 1a above). Habermas's approach to these kinds of themes almost entirely originates from an early and incomplete phase in the development of the communicative theory, prior to the publication of *Theories des kommunikative Handelns*. I shall then turn my full attention to problems linked to areas 2a and 2b, where problems pertaining to area 2b will comprise the central discussion, in accordance with the main objective of this book. It is these types of problems that are dealt with within the framework of Habermas's works from 1981 and onwards. In this connection I shall firstly (B) discuss the introduction of the distinction between the lifeworld and the system. Then (C) I shall look at the distinction between social and system integration that develops when the systems are separated from the lifeworld.[25] It is only on the more concrete level that it becomes possible to refer the relationship between lifeworld and system. In the next chapter, chapter 10, I shall be investigating whether, and if so, how, a concept of materiality can be developed within and connected to *both* the lifeworld *and* the sphere of the social systems.

A) As already became clear from Habermas's critique of Gehlen presented in chapter 4, Habermas denies that the processes of socialization in human society can be reconstructed on the basis of the natural history of the human race. Labour cannot play the decisive role in these processes. This does not mean, however, that such a history does not exist, nor that labour and technology do not play a part in it. The period in which labour occupies a crucial position in the history of the development of man, however, must come before the level we can call the establishment of society; it has to be reconstructed on the basis of principles other than man's accumulation of experiences won through his processing of nature. The same is also true for language: It has to be presupposed in order to be able to explain the emergence and development of societies. Seen in this way, Habermas does

---

primarily in Volume II, and those parts of the book which discuss more typical philosophical themes, such as the theory of speech acts and the theory of rationality in Volume I.

[25] It may seem unclear whether we should speak of system or systems in Habermas's theory. When dealing with methodological discussions it is most convenient to talk of methods adapted to lifeworld and system. When making analyses of the content of modern society, by contrast, it is useful to be able to distinguish between various social systems or systemic mechanisms.

not really need to deny the widespread theory in materialistic anthropology, that the emergence of language is linked to labour or the processing of nature. What he must deny is that the structures of linguistic communication, in the way that it forms the foundations for subsequent, and especially modern societies, can be reduced to instrumental action, to labour.

According to Habermas, the qualitatively new element in the societies of Homo sapiens is the family structure. For beings which have already mastered language, this structure renders possible three conditions for a human society, namely a perspectivist view of roles, an expanded awareness of time and connections of roles and forms of sanctions. Since language is thus already presupposed on this early level, Habermas makes room for the existence of a separate, previous level of early-human development, in which language and labour have a position as inventive patterns of development. He draws the conclusion that:

> For a number of reasons these three conditions could not be met before language was fully developed. We can assume that the developments that led to the specifically human form of reproducing life — and thus to the initial state of the social evolution — first took place in the structures of labour and language. Labour and language are older than man and society (Habermas 1979, p. 137).

He thus concludes that the role of labour for historical materialism *is* fundamental, but this is because labour plays a role as a necessary precondition for the *emergence* of societies with the division of roles and linguistic communication, and not of their form of development *when* this level is reached. He can follow Engels part of the way, insofar as labour is a component in the humanization of apes (i.e. early humans), but not in the fundamental pattern in the continued existence of mankind as socialized beings (still less as socialists). The views I have mentioned here were published in a collection of essays with the title *Zur Rekonstruktion des historischen Materialismus*. As became apparent from his critique of the subject–object model as a theory of society, however, Habermas is of course aware of the fact that the role he here allocates to labour was not *the* role allotted to labour in historical materialism.

Furthermore there are grounds for noting that the Marxist camp has also discovered the conceptual necessity of a mediation between the use of tools in animals (see also the references to Beck in Part I, chap. 2) and the fully developed human stage. The Vietnamese philosopher Tran Duc Thao raises the issue of a "most elementary form of the consciousness" in order to be able to pinpoint this identity (Thao, 1973, p. 3). He locates this in an original connection between language and labour which he calls "the indicative gesture".

When defining the gesture as guidance at a distance we have so far insisted on its form. In reality the guidance movement does not consist in simply tracing a direction, it has essentially the function of a *call*. The indicative gesture, as distance gesture, is a *call for work on the indicated object* (p. 9).

Moreover, since the gesture is a *call* it therefore implies an appeal to the other at this early stage. It already includes, Thao seems to be claiming, the reciprocity that is a precondition for all lingualistic structures.

One unavoidable problem for Habermas's simple construction is that we do not know *whether*, and if so, how, early humans could speak. In this respect we could turn to John Lyons for assistance, and presuppose the kind of distinction that he advocated between fully developed human *speech* and other forms of *lingually structured* communication (Lyons 1988, p. 147). The motor skill that we call speech is only one possible physical realization of a structure which can also be expressed in other forms of communication. On the basis of this kind of interpretation, which is doomed to remain speculative, it is possible to understand that Homo sapiens had the necessary preconditions to be able to build up family structures, regardless of the oratorical gifts of the early humans. But the question of whether, in addition to language and labour, there are also specifically biological preconditions for the human family structure, remains undecided.

The points of view put forward by Habermas that I have looked into here seem to be part of a larger, evolutionist programme, the relationship of which to the theory of communicative rationality as a whole is not entirely clear. In his later writings Habermas limited his evolutionist interest to a discussion of Kohlberg's theories on the development of the individual moral consciousness. This debate is not central to my discussion.

B) I shall now turn to the main topic of this chapter, a discussion of technology in modern society on the basis of *Theorie des kommunikativen Handelns*, which I shall be looking at in the steps I described above. I would like to remind the reader that I am pursuing a double strategy here, aiming to demonstrate both the way in which Habermas approaches technology and how his theory reveals further possibilities and further needs for an approach to technology within this framework.

Habermas's theory of modern society builds on the systematic application of a double perspective: Modern societies can be regarded as both lifeworld and system (Habermas 1987b, vol. II, p. 117 — all the following quotations from this book are from this volume). On a general level we can say that the strategic point of departure for an understanding of the position of technology in modern societies will primarily (but not exclusively) be the relationship *between* lifeworld and system, more concretely how the social systems interfere with and influence the lifeworld. It is therefore necessary to commence by giving a more exact definition of this distinction, in order to discover how it in turn affects the issue of the reproduction of *modern* societies, or, to use Habermas's vocabulary, the forms of integration — social integration and system integration — that are typical for the lifeworld and the system (or more precisely, systems) respectively.

Let us begin by considering "double" in the expression *double perspective*. In the first place it is more correct to speak of a *doubling* of the approaches we apply to society, to include *both* a lifeworld perspective *and* a system perspective. With a view to providing a critique of society it is of course important to distinguish between the lifeworld of communicative contexts and the imperatives and requirements of the social systems, but the doubling of concepts was introduced in order to enrich a social theory whose only support was its foundation in action theory. Indeed, the fundamental reason that we can talk about *the distinction* between lifeworld and system at all is the application of two different perspectives.

Habermas's decision to use both of these approaches was motivated by his confrontation with the history of sociological theory. It might therefore be useful for our purposes to describe the distinction against this background. Habermas himself mentions four reasons for his decision, all of which can be seen to be related to this confrontation. (For the following see Habermas 1991a, p. 251.)

1. The first is the historical one of two competing theoretical paradigms motivated by the wish to grasp the type of phenomenon which cannot be comprehended as attempts to fulfil the actors' intuitions.

2. He is hoping to use this reintegration to increase our understanding of those kinds of phenomena that cannot be understood (directly) by action theory. His model here is Marx's concept of real abstraction, i.e. processes that are primarily linked to the market and capital accumulation where the combined results of actions on the individual level do not only exceed, but are also incomprehensible from the perspective of the individual. The characteristic feature of systemic mechanisms is that they are uncoupled from "the members' intuitive knowledge" (Habermas 1987b, p. 149).

3. Conversely, internal problems within systems theory can be dealt with more effectively if the systems theory is applied to phenomena that have first been identified by means of action theory. In this connection he refers to problems in defining the limits and assessing the qualities of the conditions

within a social system. In our context there are grounds for making a note of this point: A systems-theoretic approach can only be applied to phenomena that have been introduced using rival terminology, in this case that of the action theory we find in the theory of communication.

4. The final reason for this double perspective is the need to overcome the problems that clung to the original distinction between the two spheres of society determined by two different forms of action, as I have already discussed in chapter 7. Habermas appears here to accept Honneth's criticism of his position from 1968, although he now claims that his new theory has in fact overcome these flaws.

No doubt the whole idea of expanding the action-theoretic approach by means of an element from a systems theory was inspired by Habermas's confrontation with Luhmann. Furthermore the background for the main concepts in the systems-theoretic tradition, which Luhmann represents, is constituted by the idea that social systems can be understood according to a pattern of self-preserving biological organisms. Habermas clearly sees nothing problematic about trying to identify theories of the market in the tradition from Smith and Marx with these basic presuppositions. In this respect he appears to be satisfied with the interpretation he had outlined earlier, that both an individualistic and a systems-theoretic methodology are equally possible and feasible as methodological approaches. They are however suited for different purposes, i.e. they grasp different aspects of society, and to differing degrees (Habermas 1970/71, p. 23). It is then perhaps most fitting to say that Habermas's use of a systems-theoretic approach does not reveal any form of ontological "commitment"; he characterizes his relationship to it as "pragmatic", in that he adopts as much of it as is necessary in order to reformulate Marx's insights in his theory of real abstraction (*Realabstraction*) (Habermas 1991a, p. 259). Systems theory can and ought to be applied primarily to those sub-divisions of society where the actors' chains of actions take on such a character of opacity that they are not understandable from the perspective of the participant and must be grasped in contra-intuitive forms of knowledge. Thus it is the area which assumes a systemic character, i.e. originally arises.

On the basis of the intense debate over the relationship between methodological individualism, which has certainly left its mark on Norwegian sociology, Habermas's attitude may perhaps seem somewhat superficial. He does not undertake any further epistemological clarification of Luhmann's basic concepts. Nevertheless, it is still possible to say that there is no necessary conflict between the double perspective used in the theory of communication and methodological individualism. As far as I can see, the main arguments for this position are not directly methodological, but rather methodological consequences are drawn from certain metaphysical, or metabiological, assumptions: Only human beings have intentions, i.e. can have plans, wishes or purposes. *This* assumption is also shared by Habermas. As we saw above under point 3, it is significant, in his opinion, that the systems-theoretic perspective can only be introduced via the action-theoretic perspective.

> Since we can gain access to the object domain in question, namely social action, only hermeneutically, *all* social phenomena must initially, irrespective of whether this is explicitly stated or not, be described in a language which takes up from the language of the actors to be found in this object domain. Given that the objectivist language of systems theory does not do this, systems analysis must rely on a different portrayal of the object domain — and here I suggest using a primarily communication-theoretic approach (Habermas 1991a, p. 254).

The division between Habermas and the methodological individualists is thus reduced to the point that he believes it is entirely possible to produce a new description of a part of society as a field of objects in terms that are not taken from action theory. The individualists support their methodological reductionism on the basis of ontological reductionism; Habermas accepts the arguments for the ontological one without rejecting a pluralism of methods. Since, however, the language of the system theory cannot be primary, but can only be introduced via that of action theory, he is able to speak of a "*methodological* primacy of the lifeworld" (p. 255, my italics). Seen from this angle, Habermas goes a long way towards meeting the individualists.

This is not the time nor the place for an in-depth study of the relationship between methodological individualism and communication theory. I would however like to mention briefly two matters: Habermas claims that the entire field of social action is only accessible to a hermeneutic approach. This will probably be hard for some theoreticians within the camp of the methodological individualists to swallow. If, however, we first distinguish between methodological individualism in general and more specific theories of the individual, such as the rational-choice theory, then no conflict arises here. Indeed, Habermas's approach seems actually to be quite compatible with the more recent trends in modern rationality theory, such as Herbert Simon's satisficing theory, which places emphasis on the fact that all rationality, also in systemic contexts, contains informal elements, mediated via traditions.

There is another important issue that remains undecided in this context: It appears that a methodological individualism necessarily implies that intentional attitudes in individuals, such as opinions, plans and purposes, are the basic concepts of sociology. Communication theory, by contrast, is a theory of intersubjectivity, the basic concepts of which are related to the consensus-forming power in the actors' illocutionary utterances. It is difficult to assess whether this entails a latent conflict. Indeed this is an area that requires further, specialized study.

These thoughts about the methodology, which are most certainly oversimplified and insufficient, bring us to another point that can be introduced by underlining the fact that we are dealing with a double *perspective* here. In reality it is the doubling of the methodological approach which is essential, as it is this that enables us to speak of *two* parts of society.

The concepts of lifeworld and system are perspectives we apply to society, in other words, for Habermas, they are first and foremost a *methodological* tool. They are not primarily, as the conceptual grid of the 1968 theory seems to have been, aimed at effecting a rigid and final division of society into spheres according to different types of action. In Habermas's critique of phenomenology he states explicitly that he wants to avoid an approach based on regional ideologies. In the final instance we can say that the concepts of lifeworld and system tend to become co-extensive, in that all social phenomena can in principle be described in the terminology of *both* theories. In practice, of course, matters are a little more complex, as we have already seen. The language of the systems theory is secondary to that of action theory, because it cannot be introduced independently. It has the advantage over the language of action theory, however, as regards its *explanatory power* with regard to such forms of social activity as capital accumulation and public administration.

> The problem of unintended action consequences can, of course, also be treated from the perspective of the lifeworld. In more complex cases, this analytical strategy soon comes up against limitations if it is meant to clarify how aggregated action consequences reciprocally stabilize one another in functional contexts and thus engender integrative effects. Such investigations must be based on a more appropriate model; and of those on offer today, that of system-environment seems to afford the greatest explanatory potential (p. 253).

It thus follows that all talk of the lifeworld and the system as two areas of society is also secondary. This way of speaking is rendered possible by the fact that those areas of the phenomenon in modern societies that can *usefully also* be described by systems theory have evolved by means of a historical process, which Habermas describes as the uncoupling (*Entkoppelung*) of the systems from the lifeworld, a process through which both the lifeworld and the sphere of the systems are subjected to an inner differentiation. This is the process I shall be investigating further in section C.

There may first however be grounds for discussing whether the theory, in the interpretation I have proposed, is in fact capable of solving the problems to which Habermas applies his theory.[26] It is perhaps the connection he postulates between action theory and systems theory and the double perspective of lifeworld and system that has resulted in most opposition from theoreticians who would otherwise feel akin to Habermas. This form of criticism makes up the most weighty contributions in the collection of articles *Kommunikatives Handeln* (Honneth & Joas 1991), and, as will have become apparent from the above, it is Habermas's responses to these criticisms that have formed the point of departure for my interpretation of this topic. All these articles criticize, albeit from different angles, the emphasis that Habermas is now placing on the systems-theoretic concept.

Herbert Schnädelbach wishes to go no further than a purely action-theoretic distinction between lifeworld and system. He locates the problem not so much in the systems theory itself, as in Habermas's concept of the lifeworld, which in his view is unphenomenological, because it breaks with the perspective of the first person. A purely action-theoretic analysis would remain within the bounds of this perspective.

> To summarize: I propose that it suffices to abide by the difference in types between purposive-rational and communicative action when developing the difference between system and lifeworld, and not to weigh this difference down with the problem of the perspectives of the first and third persons and their relation to one another ("The Transformation of Critical Theory", in Honneth & Joas 1991, p. 19).

The weakness in this proposal, in my view, is that it leads us directly back to the problems indicated by Honneth.

Hans Joas also argues against the introduction of a systems theory. He sees Habermas's motive for the introduction of this perspective as rooted in the need to be able to discuss non-intended consequences and action coordination, but maintains that this neither needs nor ought to result in the doubling which Habermas has proposed. ("The unhappy Marriage of Hermeneutics and Functionalism", in Honneth & Joas 1991, esp. p. 113.) Thomas McCarthy provides the most scathing article against Habermas, in which he contends that the introduction of systems-theoretic elements results in the abandonment of the critical points of view in his earlier writings. McCarthy is sceptical to the empirical capacity of the systems theory and its degree of realism. He points out that it has not resulted in any empirical progress in organization theory. Even formal organizations are amorphously regulated to

---

[26] I am indebted to Gunnar Skirbekk for his friendly suggestion that I explain my views on this point more precisely.

such a degree that they cannot be said to constitute any unequivocal examples of systems as defined by systems theory (McCarthy 1991). I am quoting a later, English version of his article.) His main point, however, is a political one; by perceiving politics as a subsystem Habermas undermines the possibilities for a theory of democratic will-formation (p. 169). He concludes:

> We do not need the paraphernalia of social-systems theory to identify unintended consequences. Nor do we need them to study the "functions" that an established social practice fulfils for other parts of the social network, for these are simply the recurrent consequences of this recurrent pattern of social action for those other parts (p. 177).

It ought to be obvious from the interpretation of Habermas I have outlined above that I actually share many of these doubts. I am as sceptical towards the epistemological foundations of the systems theory as McCarthy is towards its political consequences. I choose to defend Habermas on this point only because I can see a route to a purely pragmatic interpretation of the distinction between lifeworld and system. Habermas himself declares that McCarthy has shown more "essentialist connotations" of the distinction (Habermas 1991a, p. 255). Essentialist in this context is clearly meant to mean that the concepts of lifeworld and system are to be understood as directly linked to various areas of society. This is a rather strange comment, as this version must surely express the very position McCarthy himself is criticizing. Habermas does however make it clear later that even an essentialist reading will remain within the confines of a methodological interpretation, in that the subsystems are dependent on being observed and experienced by participants in order to be able to be identified (p. 255). I thus contend that it is still possible to interpret the systems concept as primarily methodological, as a means of considering specific types of action coordination, and that it thus only then can be used in contrast to the concept of the lifeworld in order to demarcate zones of society.

Gunnar Skirbekk puts forward, as far as I can see, a different type of criticism. He indicates (assuming I have not misunderstood him) that an ontology is necessary in order to be able to give the grounds for Habermas's concept of colonization. As I have understood it, he is here arguing that we must be able to speak of system and lifeworld as two parts of society in the ontological sense, what Habermas called an essentialist connotation, in order to be able develop Habermas's idea of the colonization of the lifeworld by the system (Skirbekk 1993, esp. p. 218, and note 13, p. 243 ff.). I am assuming that by ontology he means a dualistic ontology on this exact point My response to this is that if the concept of colonization were built on this kind of primarily "ontological" distinction between lifeworld and system, then it would be the concept of colonization that would have to collapse. There is quite simply no way of introducing a system concept which is "ontologically" primarily. This is, I dare to maintain, the undisputable conclusion of the last decade's critique of functionalism.

But this negative conclusion is not necessary. Habermas needs as much ontology to introduce the colonization as he does to introduce the concept of pathologies, and this is a concept which can be adequately developed on the basis of action theory. The action-theoretic concept itself provides the necessary ontology, and once we have arrived at the distinction lifeworld versus system, these two concepts can be then interpreted methodologically on this level.

There are two main arguments to support this view. The first is that the systems themselves only arise via a differentiation from an original state, which can be described as society *as* lifeworld. Admittedly, this differentiation leads to it being necessary to distinguish between two forms of integration, as I have already mentioned. System integration is however secondary to social integration since it, as I have already indicated, assumes the form of an integration of the non-intended consequences of *actions;* it thus follows that no dualistic ontology is necessary on the level of the forms of integration.

The most important argument, however, for the view which I am using as my basis, is that the rationalized lifeworld itself, independently of the imperatives of the systems, develops quasi-pathological conditions. Habermas distinguishes between impoverishment of the lifeworld via rationalization, and pathologies which develop as a result of its colonization (Habermas 1987b, p. 301, see also the next chapter). Here, then, we have a dualism as far as forms of reaction to the differentiation of systems from the lifeworld are concerned, but this dualism is clearly located within the lifeworld itself, even if the causes of the pathologies come from outside it, i.e. from the systems. Both cultural impoverishment and pathologies are forms of reaction which must be understood in terms of action theory. Thus pathologies as a result of colonization can adequately be regarded as developed via the conflict between various action structures. This discussion too results in our being able to take the distinction as purely methodological, when we confront the concepts of lifeworld and system.[27]

An innovative contribution to this debate, albeit in the same vein as the tendency in Joas, has been made by Maeve Cooke. She suggests that the dis-

---

[27] If, on the other hand, Skirbekk is merely saying that the concept of colonization presupposes a division of society into spheres, then we have more in common with one another. It nevertheless remains that no sphere can be characterized by a primarily introduced systems concept.

cussion about the usefulness of the distinction between lifeworld and system be isolated from the issue of the methodologically correct way of approaching them, i.e. from the distinction between action theory and systems theory (Cooke 1994, p. 7). The distinction between lifeworld and system thus hinges on the distinction between the two forms of integration *alone*. I would be willing to accept this for my purposes, since the theory of colonization presupposes primarily the conflict between the forms of integration. It is a different matter whether the distinction between the forms of integration alone can justify the distinction lifeworld versus system. It is interesting to note that the distinction between system integration and social integration was originally proposed by David Lockwood as an attempt at limiting the field of the systems theory (see Joas contribution in Honneth & Joas 1991, p. 111).

C) The uncoupling of the system from the lifeworld establishes in the next round the precondition for the theory of two forms of integration. It is on this level, i.e. against the background of a historical process that makes the double perspective fruitful, that it for the first time becomes possible to speak of the relationship *between* lifeworld and system. We can introduce the concept of *media* to describe the influence of the system on the lifeworld, a relationship which Habermas generally refers to as a colonization of the lifeworld (see the next chapter).

Habermas writes, as concerns the relationship between system differentiation and social integration, that:

> It is only coordinating action that harmonizes the *action orientations* of participants from mechanisms that stabilize non-intended interconnections of actions by way of functionally intermeshing *action consequences*. In one case, the integration of an action system is established by a normatively non-normative regulation of individual decisions that extends beyond the actors' consciousness (Habermas 1987b, p. 117).[28]

Elsewhere he writes revealingly:

> Thus I have proposed that we distinguish between *social integration and system integration:* The former attaches to action orientations, while the latter reaches right through them (p. 150).

---

[28] Since we are dealing with lifeworld and systems respectively here, Habermas should strictly speaking not have referred to this first case as the stabilization of a *system* of action. This must be seen as a regression to his terminology from 1968.

Habermas is now claiming that this uncoupling must be seen as a process which is nevertheless secondary, because it presupposes a prior, internal differentiation of the lifeworld itself.

> I understand social evolution as a second-order process of differentiation: System and lifeworld are differentiatied in the sense that the complexity of the one and the rationality of the other grow. But it is not only qua system and qua lifeworld that they are differentiated; they are differentiated from one another at the same time (p. 153).

This may appear to be a highly idealistic view of the historical development, a view that leaves no room for a rationalization of the economy, of the metabolism with nature. As we shall see, however, this interpretation is not entirely correct, as Habermas's point is that a differentiation of pre-modern societies, societies of a type which can usefully be perceived primarily as lifeworlds, in turn leads to a process of differentiation that releases an independent process of rationalization. Once this kind of level has been attained, the further study of differentiation of the lifeworld has to regard its material reproductions as boundary conditions. Habermas however insists that the rationalization of the inner lifeworld has its own internal dynamic. It has a "systematic effect" despite the fact that

> [Questions of *development dynamics*] can be dealt with only if we take contingent boundary conditions into account and analyze the interdependence between socio-cultural transformations and changes in material reproduction (p. 144).

The theory that the lifeworld reproduces itself via a linguistically mediated medium does not exclude, even from the point of view of this kind of purely communicative explanation, the fact that the lifeworld has a *material substratum*.

> The reproduction of society then appears to be the maintenance of the symbolic structures of the lifeworld. Problems of material reproduction are not simply filtered out of this perspective; maintenance of the material substratum of the lifeworld is a necessary condition for maintaining its symbolic structures (p. 150).

A deliberate application of Durkheim's theories has been paramount in Habermas's development of the entire theory of the processes that initiated the uncoupling of the system from the lifeworld. I am not thinking of Durkheim's holism in this context, but rather it is his idea of the transition from religious to secularized images of the world that Habermas adopts, a dissolution of the concept of the holy. As he describes the further development of this process, it assumes the definite form of the release of an internal dynamic in an independently institutionalized economic sphere. From this angle there should be no grounds for denying that his theories are at least compatible with the kinds of theories of economic modernization we find in the works of Polany and Finley. In Habermas too, the basis, in the Marxian sense, starts out as interwoven or embedded in the other social institutions, and then grows and develops to exert a pressure on them.

Against the backdrop of Habermas's general approach, however, it might be worth our while looking at this process with our point of departure in a relatively simple concept of the lifeworld and the changes it is forced to undergo on its way towards modern societies.

Habermas develops his own concept of the lifeworld through a confrontation with the theory proposed by the phenomenological sociology, an expression he uses to describe the tradition from Schütz, and later Luckmann and Berger. The materialistic tradition (Sartre) is not discussed in its own right in *Theorie des kommunikativen Handelns,* apparently because Habermas does not find this version capable of providing a satisfactory theory of intersubjctivity either. Thus the concept of materiality also disappears, although it can hardly be said of this materialistic tradition that it dissolves sociology in the sociology of knowledge and results in a culturalist overreaction, the main error which Habermas accuses Schütz of having committed.

Habermas's outline aims in the first place to overcome the obstacles of the individualistic methodology in this tradition which allows the lifeworld to be constructed from performances by the isolated individual subject. Instead the lifeworld is to be perceived communicatively. The lifeworld can in this way be seen as a network of interaction mediated by language, that is to say that it consists of individuals who are bound together by and who orient themselves in relation to the validity claims in the speech acts of other individuals. These kinds of components are included in descriptive, norm-establishing and expressive speech acts, which in turn can be traced back to prelingual components. That which, by contrast, still connects this communicative interpretation to phenomenological and hermeneutic themes, is primarily the idea of the horizon structure of knowledge. Knowledge is perceived in this first concept of the lifeworld as given in a holistically structured form, and as such, characterized by inevitable, rather than reflexive or even criticizable presuppositions.

Habermas however denies that this kind of culturalistic concept of the lifeworld is adequate for the use to which the communicative approach wants to put it in the construction of a social-scientific theory. It is a necessary to take it a step further and introduce an *everyday concept* of the lifeworld (for the problems briefly reviewed above, see Habermas 1984, esp.

chap. VI). Only now can we speak of a concept of the lifeworld that is so complex that we can refer to a *field of objects* in a social theory. The actors no longer meet only as participants in self-explanatory contexts, but also as people involved in *narratives*. This concept needs no further explanation within the scope of this work, the aim of which is to clarify the position of technology in the lifeworld and the system. Let it simply be noted that we have now reached the level of an explicit complexity of the lifeworld. From this point on, comprehension is marked by more than shared assumptions according to a horizon model, it itself gains an integrative power by serving to integrate an internally differentiated lifeworld on the basis of various aspects. It is this process that, as I mentioned above, is described with help from the theories of Durkheim and George Herbert Mead: The modern lifeworld has undergone a differentiation in three different spheres or domains: Communicative action can now serve comprehension, as well as action coordination and socialization. This corresponds to three types of processes: Cultural reproduction, social integration and socialization of individuals; and the afore-mentioned three corresponding domains of the lifeworld.

The reciprocal relationships between the domains in this dynamic need not concern us here. The point is that the differentiation between the lifeworld and the systems could only be reached as a *second* stage of differentiation on the basis of this initial and primary differentiation of the lifeworld. Nor can societies with this kind of lifeworld be perceived primarily as lifeworlds any longer, where system and social integration are linked. This first differentiation in turn results in the next, namely the uncoupling of systems from the lifeworld itself.

It is only when we have taken this step that we reach the level where subsystems of material reproduction attain such an autonomy that it is no longer reasonable to interpret them as parts of the lifeworld. Furthermore, from this point onwards social and system integration are separate and potentially in conflict with one another, a theme that is developed further in the theory of media and colonization of the lifeworld.

It is nevertheless important to underline that even this step must be introduced with the utmost care. The lifeworld and the system are not regional ontologies in the phenomenological sense, they are *not* primarily different parts of society, but rather methodological tools. To a great extent at least these concepts are co-extensive, not merely individual phenomena, but society as a whole can be regarded on the basis of both perspectives. In this way the two forms of integration with which they are associated, social integration and system integration respectively, are not restricted to previously separated areas.

In the first place, social integration itself now plays an integrating, and thus functional, role within the lifeworld. Habermas also emphasizes, especially in works after 1981, that system-integrative elements also exist *within* the lifeworld. If he had not taken this step towards a more concrete and realistic treatment, he would not have been able to overcome the flaws in the model of labour-interaction, which Axel Honneth pointed out: That it came to a halt at the contradiction between a politically neutral sphere of productive forces and an equally powerless sphere of mediation of traditions (see chap. 7).

It is only with the arrival of *this* concept of the lifeworld that this fundamental concept has been developed to the extent that we are able to query if it is fruitful to discuss the role of technology *in* the lifeworld of modern societies on the basis of the approach proposed by the theory of communication. This question and the further problem of the influence of an entirely uncoupled system *on* this kind of internally differentiated modern lifeworld are the two topics I will be discussing in the final chapter of this book.

# 10 Technology in the Lifeworld

The actual differentiation into lifeworld and systems was, as we have already seen, a secondary differentiation, which could only come about as a result of the primary internal differentiation of the lifeworld. That which *was separated*, by contrast, was the material substratum of society. This now adopts the form of subsystems, such as the autonomous market and the state government, for example. The question of whether these two phenomena can satisfactorily be understood as a further development of the metabolism with nature, and whether Habermas is capable of providing an adequate account of the relationship between the growth of market economy, the establishment of the state and bureaucratization in the modern world, will have to remain unanswered in this context, as these questions go beyond the scope of this book. Moreover there is a limit to the amount of detail we can demand in a treatment of these topics in a work like *Theorie des kommunikativen Handelns*.

Nevertheless Habermas's primary aim in this book is to demonstrate the ways in which these subsystems interfere with, obstruct and distort the lifeworld and its rationality, in the widest sense of the word, which is embedded in its communicative situations. I would like to remind the reader that Habermas's intention with the theory of systems is to reformulate Marx's original criticism based on his concept of real abstraction. *This* theory was of course intended to be more than merely descriptive; its purpose was also to describe a society that was out of control, to reveal the causes for the reified character of the capitalist society.

Although it must remain open how many of the presuppositions for and how much of the content of Marx's critique Habermas wishes to retain and develop further (Habermas 1987b, chap. VIII, p. 2), it is quite clear that the expressed adherence to Marx on this point is not without significance for the structure of his own theory. The entire construction around the double aspects of lifeworld and system must play two different roles and operate on two different levels. In the concept of the subsystems of society we find a union of both the function of the necessary material reproduction of the society and the critical idea that economy and state have taken the shape of uncoupled (*entkoppelte*) subsystems *in* the lifeworld, which are now threatening to subject it to their own patterns of system integration. The theory must therefore be able to function on both the descriptive and the normative levels, and it must be able to demonstrate the role of the systems as both differentiated parts of the original unity in tribal societies and the counterpart to the lifeworld rationality in complex modern societies. It must be able to show how material reproduction takes place in a complex and functional society. *In addition* it will also have to take on the task of indicating the role that technology plays in the subsumption of the lifeworld under the subsystems, what Habermas calls the colonization of the lifeworld.

The double role that this theory has to play is, however, a necessity for Habermas. The original rationalization of the lifeworld is after all in itself an expression of the potential rationality of modernity, for those processes it releases are the very deeply ambiguous and contradictory tendencies that are typical of modernity. This corresponds to the increasing degree of complexity in the separated systems.

> I will, therefore, analyze the connections that obtain between the increasing complexity of the system and the rationalization of the lifeworld. [...] When this tendency toward an uncoupling of system and lifeworld is depicted on the level of a systematic history of forms of mutual understanding, the irresistible irony of the world-historical process of enlightenment becomes evident: The rationalization of the lifeworld makes possible a heightening of systemic complexity, which becomes so hypertrophied that it unleashes system imperatives that burst the capacity of the lifeworld they instrumentalize (Habermas 1987b, p. 155).

The objective of this chapter is to describe the final outcome of this process, as we come across it in modern societies. We must look for possibilities for an account of technology that at the same time can function both descriptively as well as normatively and critically, and this in both a) the discussion of the lifeworld on its own, and b) the discussion of the effect of the systems on it. In the latter case we arrive at the theory of the colonization of the lifeworld. Only this last angle, and not an isolated study of the systems, will pave the way for a critical approach to the social significance of the system rationality.

As I also mentioned above, Habermas aims to round off his analysis of the distinction between lifeworld and system by introducing a concept of forms of mutual understanding (p. 187), and I shall round off my book with some comments on this concept. Habermas introduces it as a concept that, on the basis of the presuppositions in communication theory, is intended to deal with the problems concerning reification. In this way it comprises a redevelopment of Lukacs's concept of the forms of objectivity, a further development which presupposes the very criticism of the subject–object theory that crippled Lukacs's original theory.

We have therefore reached a point where we can no longer operate only with the usual distinction between lifeworld and system in general, but must now distinguish between the modern, rationalized lifeworld and those systems that are differentiated *from* and *in* it. From this point on it becomes necessary to distinguish between *two types* of media, or to use the concept of media with two meanings. The first is media in the sense of the mass media, i.e. technical media for the mediation of information and communication. This is media within the lifeworld itself. The other meaning of media is in the sense used in the tradition from Parsons, i.e. specific codes for the transfer of systemic imperatives from the systems to the lifeworld. Here Habermas operates with two media: Power and money.

In an attempt to find a common element for media in both senses of the word, media *within* the lifeworld and media *between* system and lifeworld, as I shall call them, we can say that they become necessary at that moment when the lifeworld attains such a complexity that direct transfer of communication is no longer possible, a point in its development that corresponds to the separation of the system from what was previously the material substratum. At this kind of level, threats against consensus and risks of dissent are circumnavigated by the introduction of media. The similarity between the two kinds of media must not be exaggerated however; for our purposes the distinction between them is of greater significance.

> On the basis of increasingly generalized action orientations there arises an ever denser network of interactions that do without directly normative steering and have to be coordinated in another way. To satisfy this growing need for coordination, there is either explicit communication or relief mechanisms that reduce the expenditure of communication and the risk of disagreement. In the wake of the differentiation between actions oriented to understanding and to success, two sorts of *relief mechanisms* emerge in the form of communication media that either condense or replace mutual understanding in language (p. 181).

The two types of media differ from one another in their fundamentally different relationships to semantic content; they delinguify (*entsprachlichen*) the lifeworld in qualitatively different ways. The media of the lifeworld can be taken as technical mediations of communication that do not negate the linguistic content, but which provide relief from the direct mediation of it. Modern electronic media allow mediation across expanses of space and time that are far greater than was possibly conceivable before, and the flow of information has become denser than we could previously have imagined. Media such as money and power, by contrast, provide total relief in the sense that they have replaced communicative rationality by introducing standardized special codes instead of everyday language, rendering the formation of opinions about the semantic content no longer necessary. Power and money are obviously not "alingual" in the sense that the exercising of power and the spending of money are not linked to speech, but these codes create and specify types of situations where language no longer serves as the means for achieving comprehension. In this way we are given total relief from the testing of validity claims in communicative rationality.

> The way these media function differs according to whether they focus consensus formation in language through specializing in certain aspects of validity and hierarchizing process of agreement or whether they uncouple action coordination from consensus formation in language altogether, and neutralize it with respect to the alternatives of agreement or failed agreement (p. 183).

In a different context:

> Unlike media such as money and power, they [influence and value commitment, T.K.] cannot replace ordinary language in its coordinating function, but only provide it with relief through abstraction from lifeworld complexity. In a nutshell: *media of this kind cannot technicize the lifeworld* (p. 206).

These two types of media correspond, naturally enough, to social and system integration respectively. Through the purely instrumental character which language assumes in these codes, it comes to serve only as a means in system integration, a reduction that is impossible in social integration. (Of course this does not mean that the use of power or the threat of use of power cannot be backed up by referring to legal sanctions, the legitimacy of which can in turn be debated in a discourse.) Both their functional and their dysfunctional modes of operation thus separate them from one another. While the first type at worst can only *distort* the processes of communication, the second type functions negatively by *repressing* them to a degree and in areas which are experienced as unacceptable by the members of the lifeworld. In this repression, social integration is illegitimately replaced by system integration, a replacement which entails painful consequences. Habermas calls this form of repression "colonization", a concept to which I will return below. The conflict between the lifeworld and the system has pathological consequences precisely because:

> [...] in developed capitalist societies, mechanisms of system integration encroach upon spheres of action that can fulfil their functions only under conditions of social integration (p. 305).

The lifeworld media are connected to the actual rationalization of the lifeworld, they (or rather, their negative effects) are linked to the possible degree of *loss of meaning* and *impoverishment* we find associated with the rationalization of the modern lifeworld. The Parsonian media are the basis for colonization of the lifeworld via system imperatives, and result in pathologies. Maeve Cooke deserves great merit for emphasizing the distinction between these two types of social problems so strongly (cf. Cooke 1994, and Habermas 1987b, p. 301).

These observations form the foundation for the distinction between part A) and part B) in this chapter.

A) It is then most profitable to first analyze the lifeworld separately, on the basis of the concept of an internally differentiated everyday lifeworld, at which we arrived in the previous chapter. A closer reading of Habermas's text reveals that he is not as blind to technological and material connections as he is often accused of being. Although the reproductive mechanisms in a rationalized lifeworld are also connected to communication mediated by means of symbols, it would be unfair to Habermas to interpret him as suggesting that also an internally differentiated and rationalized lifeworld enjoys a pretechnological sleeping-beauty existence. Once types of action are no longer directly correlated with specific parts of society, what Habermas calls a functionalist analysis of the lifeworld itself actually becomes possible. Let us by way of an example look at the concept of applied work force (*verräusserte Arbeitskraft*), since modern industrial labour is a field where we might expect to find that the impulses from the two forms of integration will cross and come into conflict with one another. The work force is therefore "a category" that demonstrates that the lifeworld and the system are concepts which are roughly co-extensive, as defined above.

> In this respect, the labour power sold by producers is the site of an encounter between the imperatives of system integration and those of social integration: As an *action* it belongs to the lifeworld of the producers, as a *performance* to the functional nexus of the capitalist enterprise and of the economic system as a whole (Habermas 1987b, p. 335).

Social integration, too, has functional aspects, as we noted in the previous chapter. The concept of the lifeworld at which we have arrived will now also have to be able to accommodate institutions. At this point it might be useful to remember that the distinction between lifeworld and system does not correspond to micro and macro perspectives. The rationalized and differentiated lifeworld in particular must not be confused with the micro perspective of the participants. With the rationalization of the lifeworld as our point of departure we can say that the more complex concept is reached once we *no longer* believe in the fictions that a culturalist conception of the lifeworld entails, namely that the agents are autonomous, that the cultural aspects are not subject to external forces and that the processes of comprehension are not restricted (p. 150). The concepts of pseudo-consensus and systematically distorted communication emerge on this level, and from now on we must reckon with:

> ...resistances built into the linguistic structure itself and inconspicuously restricting the scope of communication (p. 150).

It is of course true that as late as in 1981 Habermas was still writing in such a way that we can occasionally be led to believe that such phenomena are the result of the conflict with system structures *alone*. In later works, however, especially since 1984, he accentuates that the lifeworld includes all types of action, and that the lifeworld, as we have seen, also includes functionalistic mechanisms, giving us a pretext to exclude these phenomena from the lifeworld itself — we can regard them as caused by its internal rationalization. Even the socially integrated aspects of the lifeworld are no longer conceivable in the intuitive forms of knowledge of them as part of the particular culture in question.

The idea of the internally differentiated and rationalized lifeworld has thus equipped us with a concept of lifeworld with such an inner complexity that we can attribute to it all the obstructing, distorting and contrafinality-producing characteristics that are linked to empirically known processes of communication, not to mention decision-making. Of course this is what people mean when they speak of the irony in the world-historical process of rationalization. Since the rationalization of the lifeworld is inevitably linked to the complexity of the system, it also brings about an inner functionalization of the lifeworld, in addition to those external dangers which are threatening it via the pressure of the systems. Despite the fact that the process originally emanated from the rationalization of the lifeworld, it now turns on it in a process in which it appears impossible to separate the negative and positive moments from one another. Habermas believes he can establish that:

> What we have already found in the system perspective seems to be confirmed from this internal perspective: The more complex social systems become, the more provincial lifeworlds become (p. 173).

The notion of lifeworld knowledge as a resource must now not only be understood hermeneutically on the basis of the concept of background knowledge, but also economically as a means of access to necessary, but scarce resources. In his most recent major work Habermas develops this in full to its logical end.

As decision theory demonstrates, the democratic process is consumed "from within" by the shortage of functionally necessary resources; as systems theory maintains, "on the outside" it runs up against the complexity of opaque and recalcitrant functional systems, Both inside and out, the inertial moments of society — what Sartre once called the *inerte* — seemingly "become independent" (*verselbständigen*) in relation to the deliverative mode of a consciously and autonomously (*autonom*) effected sociation (Habermas 1994, p. 321).

We can therefore claim with some assurance that the lifeworld itself produces its own inertia, its own material field, through its rationalization, in addition to the field associated with the colonization of the lifeworld by the media, which is connected with technology in the more original sense of the means of production of the labour sphere. We should perhaps mention, bearing in mind that *Faktizität und Geltung*, is after all a work on the philosophy of law, that Habermas perceives the establishment of legal systems and political relations of authority as a process of differentiation of the *third order*, the emergence of a "norm-free sociality" and a "reified domain of life" (Habermas 1987b, p. 174). In the context of our analysis of a possible concept of materiality in Habermas, we can allow ourselves to ascribe less importance to this third level of differentiation.

It goes without saying that these observations are on a very general level. They have demonstrated that the communicative approach is at least not *incompatible* with a concept of materiality. On the basis of Habermas's text, however, we can also speak of a specific space for a technological mediation of the lifeworld.

We find this space in connection with a form of internal complexity in modern lifeworlds different from the ones I have dealt with hitherto, namely to the emergence of a *public sphere*. This public sphere appears in this light as a network for the transfer of forms of information, forms that in principle cannot assume the character of mediatization uncoupled from communicative rationality, as is the case with the media power and money. In cases like this (Habermas's example is influence) we are dealing with networks that are still linked to the emergence of consensus, and that "remain dependent on a rationalization of the lifeworld" (p. 183). Cognitively as well as morally recognized influence (which thus presuppose differentiated spheres of value), however, can only be transferred and mediated within the lifeworld via communication technologies.

Both kinds of influence require, in addition, technologies of communication by means of which a public sphere can develop. Communicative action can be steered through specialized influence, through such media as professional reputation and value commitment, only to the extent that communicative utterances are, in their original appearance, already embedded in a virtually present web of communicative contents far removed in space and time but accessible in principle (p. 184).

Habermas can thus also refer to the early radio surveys undertaken by the Frankfurt School and the thesis of the culture industry. (For the following see p. 389 ff., especially p. 391.) There is of course no guarantee that the mass media function optimally as far as free communication is concerned, although they cannot be totally uncoupled from the area of social integration. Their very authoritarian potential and ability for control and the centralization of the flow of information nevertheless imply that they remain within the field of communicative action. Media can, by means of various mechanisms, create both heirarchies and distortions in the horizons of communication and processes of information, but they can also wipe out the divisions within them. In principle they cannot therefore be delimited from or negate the need for the individual formation of opinions, despite the fact that they can cripple it. The result of the dissolution of local frames of interpretation and comprehension does not have any specific tendency per se, be it in a negative or a positive direction.

Habermas thus comes to a standstill at the not particularly original thesis of the ambivalent nature of the mass media, in opposition to the predominantly negative interpretation of modern media in the early Frankfurt School. This ambivalence is, in Habermas's opinion, an inevitable property of the media. In the same way that there is no guarantee against distorted communication, there is something which can uncouple the media once and for all from a potentially critical opinion formation to (and in) them.

There is another aspect that is of greater significance to our perspective than these purely theoretical discussions about media. It seems that a modern, rationalized lifeworld with mediatized public realms cannot be conceived of without a concept of materiality, which is made up by the media themselves in this sense. We can, however, also see how this concept will assume a different position to that which it occupies in social theories based on the theory of consciousness. This field will not arise through the constitution of the society by the individual or the collective subject. It will be a field or network which, positively or negatively, mediates the linguistic interactions that are the medium of social integration.

Communication technologies can determine the form of linguistic mediation; exchanges of information are steered by media presuppose a communication that is at the same time both generalized (uncoupled from specific spatio-temporal contexts) *and* specialized. Habermas speaks of technical expansion, organizational mediation and rationalization of comprehension (p. 267). They are not capable, however, of shattering the basic principle within the theory of the lifeworld that language is a medium for understanding.

This is probably as close as we will come to finding a positive connection between Habermas's concept of communicative rationality and a concept of materiality. (The use of the concept of materiality linked to media in the second meaning of the word, which in fact comprises the typical and traditional themes in the criticism of technology, will be dealt with in part B of this chapter.) Of course this need not imply that we are reverting to labour as an archetype for reflection. A theory of social integration that builds on the binding power in illocutionary utterances and the formation of opinions by everyone, qua ego, to the utterances of the others, qua alter, also presupposes frames for interpretation of speech acts. On the most general level it is possible to say that the lifeworld forms a potential resource for both interpretation of and formation of opinions about speech acts. It is these universal *frames* for these interpretations that must be said to be technologically mediated to a crucial degree in a mediatized lifeworld.

Writing, the printing press and the electronic media mark the significant innovations in this area; by these means speech acts are freed from spatio-temporal contextual limitations and made available for multiple and future contexts... [the printing press] brought with it a freeing of communicative action from its original contexts, this was raised again to a higher power by the electronic media of mass communication developed in the twentieth century (p. 184).

The implicit presuppositions about relevant contexts, which we make in all interpretations, can no longer be studied independently of the technical mediation of the speech acts. In public spheres they only reach us filtered through the material field of the communication media. This has a reciprocal significance for the relationship between the perspectives of first and the third person. On the one hand we have technical conditions that allow communication of semantic content. It might still be possible to claim that we only have primary access to these kinds of relations through a hermeneutic approach. Nevertheless it is clear that these conditions can only be understood fully in a third-person perspective. On the other hand, this makes also the material field relevant for the very validity claims of the alter about which we can only form an opinion in a first-person perspective. We can no longer regard a concept of materiality or a concept of technology merely from a third-person perspective, as an anonymous process that happens to us. This is of course not *only*, but obviously *also* a structural change. If speech acts in general have to be perceived as technologically mediated, then technology is something we meet, if not directly consider, also in a first-person attitude. It is only from this angle that we are first able to interpret, and then to form an opinion about the validity claims in speech acts of alter. The distinction between the two types of perspective is not annihilated, but they cut through one another in an extraordinary way, which seems to be peculiar to modern lifeworlds with mediatized public spheres.

Finally, then, it is worth taking the time to see how the general theory of a lifeworld looks from the perspective of this concept of a technically mediated modern world.

In a relatively late work Habermas develops the concept of the lifeworld further, especially with the concept of lifeworld *knowledge* in mind ("Handlung, Sprechakt, sprachlich vermittelte Interaktion und Lebenswelt" in Habermas 1988. For the following see in particular p. 88 ff.).

In the first instance Habermas distinguishes between a situation-specific background knowledge and a topic-dependent contextual knowledge. We can take these two types as reservoirs of knowledge that can be mobilized both in the interpretation of speech acts and in the formation of opinions about the validity claims that are raised, e.g. in the issue of the relevance of what has been said. (For a very enlightening discussion of the relationship between relevance and Habermas's own list of validity claims, see Cooke 1994, chap. 3.7.) It does not take a vast amount of imagination to see that technology will always comprise an element here not only in the transfer, but also in the evaluation of the conditions of acceptability. This becomes apparent in Habermas's own example: The statement "It is snowing in California" said in a park in Frankfurt is acceptable if the person who utters it has just returned from California or is a weather forecaster. The conditions of acceptability are thus immediately seen in the light of the fact that it is possible for a traveller who is in Germany today to have first-hand knowledge of the weather in California, or that it is possible to have technically mediated expert knowledge of it.

There is, however, another issue which is more interesting in this context. Habermas contends (Habermas 1988, p. 90) on the basis of the work of John Searle, among others, that both the forms of knowledge mentioned above are rooted in a background knowledge of the lifeworld. This kind of knowledge is implicit and not intentionally assumed in all evaluations of the validity of speech acts, or even in the understanding of their meaning. For example, in order to understand of the statement "The cat is on the mat" it is *presupposed* that gravity is working in the way it usually does on the face of the planet Earth. It is not difficult to imagine that a thoroughly mediated everyday lifeworld, dependent on modern technology in all areas, could quickly and easily result in changes in our spontaneous perception of which constants are normally relevant for the conditions of validity. In this context we can only give a brief summary of the main elements of these ideas. Habermas also writes of the example of gravity:

> Since Homo sapiens started to sustain his life by the use of certain tools, he must have had an intuitive knowledge of the principle of levels (*Hebelgesetz*). But it was modern science which just discovered it as a law and transferred it into the form of explicit knowledge by a methodic investigation of own pre-theoretical knowledge (Habermas 1988, p. 90).

The point here is not the direct relevance of the distinction between background and foreground knowledge, but rather that what functions prereflexively in the modern world is by no means original or independent of theory in the sense presupposed here. By means of a "secondary traditionalization", elements of experience and contact with advanced technology (which does not imply any theoretical knowledge of the apparatus on the part of the *user*) will become an integral part of the background knowledge of humans in the modern world. Habermas falls prey here to a preconception which is common in the phenomenological theory from Heidegger to Merleau-Ponty, namely the misconception that whatever seems prereflexive to the individual in an action situation is also originally given in a genetic sense. Seen in relation to how deeply rooted the use of tools is in the human physiology and the entire history of the development of the species, this is not an unreasonable view of labour on the level of a craft or similar skills. But as soon as we start looking at modern, industrial labour, where the basic structure of the operations performed by the workers is determined by class-specific, financial motives, however, this type of analysis seems to have less relevance. I am not trying to suggest that Heidegger should have become a sociologist instead of a philosopher, but rather I am rejecting the claim that at this level, what Heidegger called "to hand" (*zu-Handen-heit*), is fundamental to modern man in the way he suggests. Nor can the tool, the use of tools or similar skills continue to serve as paradigms for the way in which we relate to artefacts (Heidegger 1927, § 15).

Our existence today can no longer be described adequately by this phenomenological view. Theory no longer, or at least no longer only, reveals and reformulates that which makes up our background knowledge, but rather theory is now itself a presupposition for it. Our background knowledge and the machine technology are not related to one another in the same way as the lever and classical statics, i.e. as a before-and-after relationship.

This is the result of living in a lifeworld that is also marked by its being a field of materiality, and of precisely that materiality which is produced by modern technology in all its different guises. The salient point here is the significance this has for the character of such a horizontally and holistically organized knowledge. It seems at first glance that my view implies that our lifeworld is more open to reflection than earlier ones.

It might be useful at this point to summarize these points of view. They do not directly affect the actual theory of the communicative rationality and the orientation towards validity claims as the binding agent of the lifeworld, as I have already explained.

Nor is it my view, if we now descend to the more specific level of a theory of the lifeworld, to be taken as a criticism of the analytic distinction between culture, society and individual. There is however one point on which the more traditional point of view must be revised.

This point is that not only cultural traditions (as Habermas asserts in Habermas 1988, p. 98), but also maintenance of the structures of society and of interpersonal relationships — which play a part in constituting personality — are mediated via a world of objects, via materiality.

We have thus arrived at the fundamental difference between this concept of materiality, associated with the media of the public sphere, and that which we will reach in section B). The material field of social integration is inextricably linked to the communicative mechanisms of the lifeworld. Integration takes place here through the attitudes adopted by participants to each others' illocutionary utterances, which in turn means forming opinions about whether something true is said, valid or expressively authentic, and arranging their plans of action accordingly. Technology will condition at best only the frame of interpretations for them; media, as *the* binding agent between system and lifeworld, has no such relationship to semantic content and will be of very different significance for lifeworld relations.

B) The next and final stage deals with the position technology occupies in the relationship between the lifeworld and the differentiated subsystems within it. This position can be approached from both 1). the point of view of the system, where we come into direct confrontation with the concept of media in what I called the second sense above, and 2). from an action-theoretic examination of the system rationality. I will discuss these two approaches in this order.

1) The duality I indicated earlier, the necessity to describe both the necessary function as well as the colonizing characteristics of the system, and then establish a descriptive as well as a critical and normative perspective, becomes particularly pressing here.

These two contrary tendencies clearly mark a polarization between two types of action-coordinating mechanisms and an extensive uncoupling of system integration and social integration. In subsystems differentiated out via steering media, systemic mechanisms create their own, norm-free social structures jutting out from the lifeworld. [...] We cannot directly infer from the mere fact that system and social integration have been largely uncoupled to linear dependency in one direction or the other. Both are conceivable: The institutions that anchor steering mechanisms such as power and money in the lifeworld could serve as a channel *either* for the influence of the lifeworld on formally organized domains of action *or*, conversely, for the influence of the system on communicatively structured contexts of action. In the one case, they function as an institutional framework that subjects system maintenance to the normative restrictions of the lifeworld, in the other, as a base that subordinates the lifeworld to the systemic constrains of material reproduction and thereby "mediatizes" it (Habermas 1987b, p. 185).

Habermas is thus operating with two media (money and power) as opposed to Luhmann, who has four (money, power, truth and love). Of these four, Habermas clearly believes that truth and love cannot be differentiated and mediatized, i.e. function in a purely systems-integrative way, independently of semantic content. At this point it is necessary to underline once more the inadvisability of interpreting these concepts according to a topological model, or according to a pattern of a phenomenological theory of regional ontologies. It might be useful here to look at what Luhmann has to say about power:

All media of communication are, if at all differentiated, in this way arrangements of society as a whole; truth, money and even love are in this sense omnipresent, and it is a necessity of existence to participate in them in a positive or a negative way (Luhmann 1988, p.90).

For Luhmann, who is basing his model on a consistently systems-theoretic conception, all four media can be assumed to have basic common features in their codification (despite specific differences among them). In Habermas's system, by contrast, things have to be different. The two media that can really be differentiated from the lifeworld and linked to subsystems such as the market and the state, must have a fundamentally different relationship to language than that we found in the modern forms of the public sphere, where communication is mediated via modern mass media. In relation to alter as a speaker, there is always a moment of acknowledgement, which consists of interpreting the statement of the other party as a potentially acceptable validity claim. This does not occur in those relationships that Marx called real abstraction and which, as we have already seen, were Habermas's point of departure for his assumption that certain relationships in society can best be analyzed using systems theory. Thus interactions are no longer:

> [...] coordinated via norms and values, or via processes of reaching understanding, but via the medium of exchange value. In this case, participants are primarily interested in the consequences of their actions. Inasmuch as they orient themselves to "values" in a purposive-rational manner, as if the latter were objects in a second nature, they adopt an objectivating attitude to each other and to themselves, and they transform social and intrapsychic relations into instrumental relations (Habermas 1987b, p. 336).

As we have observed, the two media in question here provide a qualitatively higher degree of relief from communication than is possible within the lifeworld; through them an almost complete replacement is found for an interpretation of a semantic content. The designation "systemic" seems thus to indicate a marginal value. This becomes clear if we compare media like power and money, with media such as influence and investment of assets which, as we saw, cannot be uncoupled from the lifeworld.

The concept of technique which is used here seems to be borrowed from Luhmann, who defines it as follows, in conditional support of Husserl (for the latter, see (Blumenberg 1963):

> We consider the essence of the technical (*des Technischen*) to be ... the relief of the processes of experience and action, which process meaning, from having to absorb, formulate and make explicit in communication all the implied elements of meaning.....This concept of technology is much more comprehensive than the concept of machine technology...Thus it has an extension which is adequate for the system of society as a whole (Luhmann 1988, p. 71).

It is worth noting that Luhmann refers to a technization of media, whereas Habermas ascribes to him a theory of the technization of the lifeworld. Regardless of this, there can be no doubt that this quotation serves to support the thesis that total technization means relief from all linguistic communication, or, more precisely, *would* mean that if it were possible. In Habermas's view this is an unattainable point, a veritable *vision* of horror. As in his early writings, Habermas insists that a complete technization or technocratization of society is impossible, and that the belief that this sort of thing is possible is an ideology which disguises the fact that these processes have gone too far. In developed capitalist societies, mechanisms of system integration threaten fields of action, which can only fulfil their function under the conditions of social integration.

We have now reached the position for and formulation of the classic criticism of technology within Habermas's theory of communicative action. Communicative contexts linked to social integration cannot be replaced, but they can be, and indeed are, obstructed and crippled. They do not disappear, but they cannot fulfil their function when a cognitive, instrumental rationality is allowed to control their territory. The essential point with regard to social criticism is that it becomes possible to separate on the one hand *that* threat against the lifeworld which forms the foundations for the distortion of normative rationality and aesthetic subjectivity, and which is caused by the internal dynamics in the pressure from the state and the market via the media power and money, from the conservative topos of loss of meaning and mass society on the other.

We can summarize these thoughts with help from a quotation which at the same time also expresses the critical potential in the media analysis.

> Things are different when system integration intervenes in the very forms of social integration. In this case, too, we have to do with latent functional interconnections, but the subjective inconspicuousness of systemic constraints that *instrumentalize* a communicatively structured lifeworld takes on the character of deception, of objectively false consciousness. [...] This gives rise to a *structural violence* that, without becoming manifest as such, takes hold of the forms of intersubjectivity of possible understanding (Habermas 1987b, p. 187).

One can nevertheless wonder if this concept of technique is sufficient. It accentuates a particular mode of operation in all technology, namely its relieving quality. This can accommodate the phenomenon that Habermas called the irony of rationalization: That this kind of relief goes hand in hand with, and even causes increased complexity and differentiation of systems. As a result of the conflict between the technization tendencies, which are linked to money and power, and the communicative rationality of the lifeworld, it becomes clear that this kind of tendency both is anthropologically necessary and at the same time that it is incapable of covering all areas of human life. In this way it can be said to accommodate both a descriptive and a critical element. It is doubtful, however, whether the theory is complex enough to illustrate the specific power structures in modern societies and the role of technology in maintaining and preserving them. It would go beyond the scope of this book to assess whether Habermas underestimates the class conflicts in modern societies (see p. 340).

Irrespective of which types of power structure we operate with, the concept of technique that Habermas introduces here can advantageously be supplemented with a concept of materiality. In the first place, this concept can explain how the systemic imperatives *become embedded* in the lifeworld itself, how the impact from those decisions mediated by the media from the market forces and government can operate immanently in and through the pores of a society structured on the basis of a completely different rationality.

The only other researcher I know of who has tried to introduce technology in connection with Habermas's theory of the media, is Andrew Feenberg (Feenberg 1994a and Feenberg 1994b). I hope my comments on his approach will also help to shed some light on my own interpretation, and that the criticisms I make will not overshadow the fact that we share many of the same views. We both believe that the development of technology in modern societies both expresses and cements the power structure, what he calls the theory of implementation. The issue at stake is how these insights can best be expressed on a theoretical level.

Feenberg's proposal is quite simply that we should perceive technology as a separate third medium, in addition to power and money (Feenberg 1994b, p. 11), while I would prefer to introduce the concept of materiality as a mechanism of mediation for these two media. He has produced his own version of Habermas's diagram no. 37 (in Habermas 1987b, p. 274) to demonstrate how this can be achieved. My critical comments concern both the extent to which this proposal is practicable as a theory of media seen in isolation, and the actual concept of technique that it presupposes.

As Feenberg himself says (Feenberg 1994b p. 15), his theory assumes that the goal is to develop an ideal-typical account of technology, and I agree with him that merely to say that in practice all media will mediate one another does not constitute a legitimate counterargument. Furthermore it is not implausible to claim, as Feenberg does (Feenberg 1994b, p. 12 ff.), that it is possible to observe a certain affinity between the functions of the media with regard to providing relief from risks of dissent in communication on the one hand and the general role of technology as an agent of relief on the other. The media entail a legitimate delinguification, within their boundaries, of messages; they relieve the recipient, albeit implicitly, from forming opinions about the validity claims in the sender's speech acts. Even in legitimate budgetary decisions made by the government, money and power will mediate one another, and since this is not an acceptable argument against regarding money and power as separate media, the same must then also hold true for technology. It obviously goes without saying, however, that this is far from enough to *prove* that technology is indeed a separate medium.

Furthermore, other objections surface if we take a look at Habermas's discussion about the proposal to make influence and investment of assets separate media, in the same way as Feenberg hopes to do with technology

(Habermas 1987b, p. 273). In the first place media would have to include a special substance which can be calculated, stored and externalized. There is nothing of this nature in technology, with the possible exception of energy and this is obviously an unfortunate analogy.

In the second place technology is not bound to one set of normatively sanctioned and regulated positions and institutions, such as we find in bureaucracy and the legal system. The position engineers have acquired is due to their place in a production process that is regulated by the private property of the means of production. (The unhappy fate of the American technocracy movement is a clear expression of this.)[29]

Moreover I am sceptical to the concept of technique which Feenberg implies in his extended scheme. As a real value he quotes the realization of goals, and as a reserve, natural consequences. With this kind of concept of technology the social element suddenly disappears. In the first case we appear to have returned to technology as a means for the realization of objectives. In the second, technology appears to be the control and correction of failed attempts to master nature. The idea that technologies can be scrapped on other grounds than that they are inconsistent with the laws of nature, that they are systematically selected on the basis of profit, and not on the basis of a completely abstract criterion of performance is lost. On the ideal-typical level, Feenberg's concept of technology is socially neutral. I do not see how any ideal-typical concept of technology can merge with Feenberg's anti-essentialism on this point.

Of course Feenberg did not set out to defend this sort of concept of technology. I attribute his ending up in this impasse to an unconscious inheritance from Marcuse. In Marcuse's writings we find a wavering between a concept of a technology with a utopian potential for change and a technology that is comprised in its entirety of means for control and domination. Feenberg does not have these utopian pretensions, but in his work it is also unclear *on which level* politics and technology are linked: If it is on the ideal-typical level of the abstract concept of technology as a medium, then we need a different concept of technology to the one implicitly introduced. If however it is only in the concrete manifestation of technology, where power and money are already mediated and we can find traces of power constellations, then no concept of technology as a separate medium is necessary.

---

[29] In his scheme Feenberg defines systems as a form of institution for technology. We are well aware today that technology is organized in systems, and cannot be regarded as isolated discoveries or machines. We are not, however, dealing with subsystems of society here, such as the legal system or public administration, and thus we remain without a parallel to the institutionalization of the media.

These matters are treated satisfactorily in his thesis of implementation, which is easily reconciled with a concept of materiality.

Once we regard this argument against the background of the internal problems of defining technology as a medium, I think I see good reasons for sticking with the idea of technology as materiality, as a mechanism of mediation for systemic imperatives, where economic and political decisions are embedded, solidified and reproduce themselves. In this way we avoid the highly problematic conception of a technology which is supposed to be biased, but with no further foundation — a concept which I find highly dubious (Feenberg 1994a, p. 98). That technology in its concrete form reveals a bias for the prevailing conditions can be explained satisfactorily by the theory of materiality. This is true for technology in connection with both types of media. Winner's example of Robert Moses's bridges, which were too low for any buses to be able to pass underneath (Feenberg 1994b, p. 11), seems to be cut out for an analysis in terms of the theory of materiality. In these kinds of analysis, the concepts of implementation and materiality indeed dovetail quite nicely.

And in reality the difference between Feenberg and myself is more complicated and indeed paradoxical than my discussion so far has revealed. I did indeed claim above that the theory of technology as a separate media burdened him with an unintended ahistorical concept of technology. But since I am more in favour than Feenberg of a possible instrumental core logic of technology in terms of relief, one might suggest that I should be more open to a media concept of technology than him!

Indeed apart from the specific critique above, in one sense I am. Feenberg's claim that Habermas has overlooked technology in connection with media theory, and in the whole of *Theorie des kommunikativen Handelns* is the crux of the matter. I would like to claim that the opposite is the case; the whole theory of media formation — perhaps the whole of volume two of this book — is nothing but a theory of technology, technology taken in the wide sense we found in Luhmann above. The difference between a Luhmannian and a Habermasian theory would then be not so much the sense of the term technology, but its extension. *This* disagreement has, in its turn, deep roots. Habermas was of course inspired by Luhmann to adopt from Parsons the concept of media, but his reduction of Luhmann's four media to two, dropping truth and love, is inspired by his original intuition, which developed into the distinction between labour and interaction. Luhmann in his turn retaliated by saying that the distinction between practice and technique was antiquated, and regarded this point as a significant difference between himself and Habermas (See Luhmann's article "Systemtheoretische

Argumentationen. Einen Entgegnung auf Jürgen Habermas" in Luhmann & Habermas 1971, p. 293, p. 377).

So the basic concept is technology, and media formation is secondary to it. The formations of specific media are concrete occurrences of the process of technization, where parts of the field of communication become independent from the normal linguistic process and take on the forms of specific codes. This process may then, so to speak, turn round and bite the hand that feeds it. This may have some analogy to what Georg Simmel called the tragedy of culture, when the inventions of human culture become reified and take on a life of their own, opposed to the will of their inventors. Of course this concept lacks the force to deduce the specific traits of different types of media formation, as in the case of money.

So Feenberg was quite right in spotting the *media-like* features of technology, but I would like to offer these considerations as a possible reinterpretation of the order of priority between technology and money and power as media. And on this account, it is no mystery that specific forms of technology lack the power to become fullblown independent media in their own right.

2) Hitherto I have looked at the relationship between system and lifeworld on the basis of a systems-theoretic approach. We can however equally well apply the opposite perspective. In the same way that we saw that social integration has certain functional aspects, there are also grounds for drawing the conclusion, from what I observed in the previous chapter, that the systems too can be subjected to an action-theoretic analysis. I will deal with this problem first, before moving on to Habermas's theories of media as the connection between system and lifeworld, an area best approached from the systems-theoretic angle.

It is important to view the systems even in the light of action theory for two reasons: Firstly it confirms our intuition that these two perspectives tend to be co-extensive, although action theory is, as we have seen, methodologically primary. Secondly, only the implementation of this kind of analysis can confirm that Habermas is no longer caught up in the dualism, which Honneth criticizes, between a purely normatively regulated and a norm-free part of society. In 1981 Habermas had yet to draw the full consequences of this. In a later work, however, he writes of the forms of action in the subsystems:

> The acting subject is able to retain his/her success-oriented and in borderline cases [...] purposive-rational stance, but only under conditions of an *objective inversion of the ends set* and *the means* chosen, for the medium itself is now the transmitter of the respective subsystem's system-maintaining imperatives. This inversion of means and ends is

experienced in the form of the reifying character of objectified social processes. I would thus miss the point if I were to continue to speak, as I did in earlier publications, of systems of purposive-rational action. Media-guided interactions no longer embody an instrumental, but rather a functional form of reason (Habermas 1991, p. 258).

Habermas is referring to his own works from the 1960's here.

The first point is that this angle to the problem transports us beyond the rigid dichotomy in the distinction between labour and interaction. A second is that this improvement makes his action theory more capable of handling modern technological systems than the simpler concept of purposive-rational action. In this way Habermas is moving along the same tracks as other modern technology theoreticians who all emphasize the fact that modern technology cannot be apprehended as the means to predetermined ends, but that it itself circumscribes the total sum of the objectives to be realized. It is precisely these sorts of aspects of contact and interaction with technology that are inherent in Habermas's conception that certain parts of society assume a character which can best be described as systemic.

One problem remains, however: How can we in a single theory unite the colonizing side effects of systemic intervention with the *necessary* forms of the technological imperatives from the material reproduction of society? How can we distinguish between the colonizing and the necessary imperatives, which are after all both mediated by the same material? At this point Habermas introduces the concept of *forms of mutual understanding* (*Verständigungsformen*), established on the basis of Lukacs's concept of forms of objectivity. In Habermas's version we are dealing with formal qualities of forms of understanding, as opposed to conditions for possible experience of objects. We cannot really speak of more than a guiding principle in this context, but nevertheless we are faced here with a possibility for marking the distinction between legitimate and illegitimate zones of technological imperatives.

> A form of mutual understanding represents a compromise between the general structures of communicative action and reproductive constraints unavailable as themes within a given lifeworld. Historically variable forms of understanding are, as it were, the sectional planes that result when systemic constraints of material reproduction inconspicuously intervene in the forms of social integration and thereby mediatize the lifeworld (Habermas 1987b, p. 187).

Habermas develops this thought in connection with an evolutionary theory of the history of the distinction between lifeworld and system. He distinguishes first between four fields of practice: practice of religious cults, practice—

governing world pictures, communication and purposive activity (p. 190). These are then combined with four levels of cognitive differentiation.

Habermas does not attempt to develop this scheme further than to apply to early-modern societies, which are characterized by the differentiation of a specific and not merely implicit problematization of validity claims. (For this concept which was first developed in Habermas 1972, see Part III, chap. 1 above.) Here the field for purposive action is indeed characterized by a value-free exploitation of purposive-rational action (p. 191).

Admittedly this concept is scarcely given the position which Habermas seems to promise (p. 110) in the final version of his new theory of rationalization (chap. VII). Nor is an analysis of technology beyond the technology of the mass media included in his list of the tasks which we can expect a critical theory of society to fulfil (p. 383), and we can also find fault with the fact that the shape that purposiveness takes in the scheme sketched above leaves us with an aftertaste of the value neutrality that haunted the distinction between labour and interaction.

I shall therefore try to demonstrate that there nevertheless is potential in the idea of forms of mutual understanding as a designation for the fields of the conflicts between lifeworld and system imperatives.

\*

In the Introduction I stated that I would conclude this book with a discussion of Habermas's theory because I regarded it as the most promising theoretical approach to technology today. I also mentioned that we would find it necessary to fall back on interdisciplinary research strategies and that I basically doubted the existence of anything that might merit the title *the* theory of technology. I am therefore not claiming that the communicative theory is capable of forming the framework for such an all-encompassing theory. First of all I would like to indicate the areas to which it is unable to gain access.

Apart from the idea of functionalist reason or form of action, the communicative theory does not appear to make any noteworthy original analyses of the systems, seen in isolation. We lose the purely engineering-scientific elements, which, after all, are necessary for an explanation of why nature can be shaped in one way and not in another, and why the social materiality that thus emerges, guides our relations in this way, rather than that. Elements typical of a technological path of development, such as bottlenecks, blockages and the pressure, once these problems have been surmounted, which arises in the interface between economic and technological decisions, are also ignored. The theory tells us little about the interplay between political and technological decisions, and between state bureaucracy and technology in large concerns. Here at any rate, however, the double conception of media, which includes both money and power, comprises a plausible approach.

The area in which I find this theory unsurpassed, however, is its ability to grasp the interplay, which is full of tensions, between technology and other functioning forms of human activity (the systems) and those forms of the elements of a society that cannot be reduced to a functionalist perspective. What we are lacking is a further concept capable of showing how the systemic imperatives are able to fasten their grip. And I find this in the concept of materiality. It supplements the concept of technization by showing how media are mediated to and within lifeworld situations. This means that the interactions and forms of operation where a relief from an interpretation of semantic content, i.e. an integration by means of media that allow mediation via objectivated entities only, is no longer possible in principle. If Habermas's theory as it stands is capable of formulating a concept of the world-historical irony in the rationalization, a materiality-theoretic extension would also enable the theory to explain in detail, not primarily how this irony first arose, but the manner in which it *operates*. In this light I find grounds for keeping Habermas's concept of forms of comprehension, in its sense of zones of conflict where the collision becomes acute between such principles of integration as are reciprocally irreducible, but which also complement each other and are equally necessary for the existence of society.

Seen in this way, the double perspective on society (perceiving society as both lifeworld and system simultaneously) can *also* shed a new light on the systemic character of society. For this too is now seen in a double perspective. The systemic nature of society cannot only be judged to be negative. It is after all this that enables technology to be thought of as controllable. The alternative to living with technology as an ever-present, problematic political decision is a pre-modern state where system and social integration merge in such an inconspicuous way that society really is left open to anonymous forces.

> It's only at relatively high levels of internal dependency that it is possible to manage social systems. It is, for example, not strong conservative traditions, insuperable institutions and close-knit social systems that make the Indian economy so difficult to regulate. It is rather, as Gunnar Myrdal pointed out in his now rather dated Asian Drama, the softness of the Indian economy, the far too low level of general systemacy in its basic technological operations. This lack of systemacy makes it extremely difficult to regulate. [...] Impulses of change introduced simply die out as the parts of the system are too loosely knit (Beckman 1990, p. 28).

In Habermas's later works we also see a tendency to reject earlier ideas that the state and the economy will dissolve society, and thus be completely subjected to political control (See for example "Vorwort zur Neuauflage", in Habermas 1990c, p. 33 ff.). Bureaucracy and technology are phenomena that have a momentum which ought to be harnessed and mastered rather than negated, and which have their own value and their own legitimacy. This set of problems is inextricably connected to what it means to live in a modern society.

The problem with technology is therefore the tendency of modern societies to generalize the systemic features, to fail to limit the systemic imperatives, rather than the tendency to develop subsystems seen in isolation.

The advantage of the double perspective that the communicative theory offers us is expressed here in two different connections.

The first is its ability to distinguish between two different relationships: On the one hand the inner differentiation of the lifeworld and the differentiation, rationalization and normalization of the modern rationality that this entails, e.g. in the emergence of modern art and science; and on the other hand, the cultural impoverishment that can occur when communicative rationality is mediatized. And this must in turn be distinguished from the repression of communicative rationality through a technization. The sum of these distinctions enables us at least to see that there is a difference between the content of that which is genuinely modern and the irony, or dialectic if you prefer, of modernization.

The second connection that demonstrates the usefulness of this double perspective is that this theory allows us at least an insight into the kind of elements we would need in a more precise description of what happens in the zone of conflict between the lifeworld and technology. The concept of forms of mutual understanding is interpreted as a step in this direction. We can understand reification as the experience of the fact that a symbolically mediated interaction, which in turn must be technologically mediated, is repressed by forms of functionalization where the objective medium itself, and not the attitude towards alter's illocutionary acts, constitutes the integrating force. This makes it necessary for us, as I have done in this chapter, to distinguish between materiality as a bearer of communication and as a bearer of systemic imperatives. In this way the communicative theory not only allows, but actually forces upon us an important further development and differentiation of the concept of materiality in relation to both the Marxist and the phenomenological traditions.

As I have already pointed out, we do not come across a complete theory in Habermas's writings. The essence of its significance today lies in the fact that it enables us to make some essential distinctions that can pave the way for further analyses.

Technology is not identical to politics — it has its own internal dynamics. Nor is it culture, or even the modern lifeworld itself. Perhaps however the theory of the double perspective can at least indicate the correct position for a concept of materiality and of the social meaning of technology: Somewhere between the detached registration of the systematic character of all social reality and the paranoid reverse side of this point of view — that everything is power.

# Bibliography

Adorno, Th. et al. 1976. *The Positivist Dispute in German Sociology.* London.
Armstrong-Kelley, G. 1973. "Bemerkungen zu Hegels 'Herrschaft und Knechtschaft". In H. F. Fulda & D. Henrich (eds.), *Materialien zu Hegels "Phänomenologie des Geistes".* Frankfurt.
Aronson, R. 1980. *Jean-Paul Sartre. Philosophy in the World.* London.
Ayers, M. A. 1970. *Perception and Action.* In: G. A. N. Vesey (ed.), Knowledge and Necessity. London.
Babbage, C. 1832. *On the Economy of Machinery and Manufactures.* London.
Balibar, E. & L. Althusser. 1970. *Reading Capital.* London.
Baruzzi, A. 1973. *Mensch und Maschine.* München.
Beck, B. B. 1980. *Animal Tool Behaviour.* New York.
Beckman, S. 1990. *System and Systemic in Technology* (Working Paper No. 20, Center for Technology and Culture). Oslo.
Beckmann, J. 1780. *Anleitung zur Technologie.* Leipzig.
Benjamin, W. 1936."Das Kunstwerk im Zeitalter seiner technischen Reprodusierbarkeit". *Zeitschrift für Sozialforschung*, vol. 6.
Berg, M. 1980. *The Machinery Question and the Making of Political Economy 1815-1848.* Cambridge.
Bijker, W. E., T. P. Hughes et al. (eds.). 1989. *The social Construction of Technological Systems.* Camb., Mass.
Bloch, M. 1983. *Marxism and Anthropology.* Oxford.
Blumenberg, H. 1963. *Lebenswelt und Technisierung unter Aspekten der Phänomenologie.* Torino.
Blumenberg, H. 1973. *Säkularisierung und Selbstbehauptung.* Frankfurt.
Blumenberg, H. 1976. "Selbsterhaltung und Beharrung. Zur Konstitution der neuzeitlichen Rationaliät". In H. Ebeling (ed.), *Subjektivität und Selbsterhaltung.* Frankfurt.
Borgmann, A. 1984. *Technology and the Character of Contemporary Life.* Chicago.
Braudel, F. 1972. *The Mediterranean and the Mediterranean World in the Age of Phillip II.* London.
Bruland, K. 1989. "The Transformation of Work in the European Industrialisation". In P. Mathias & J. Davis (eds.), *The first industrial Revolution.* Oxford.

Bruland, T. 1982. "Industrial Conflict as a Source of Technological Innovation". *Economy and Society*, vol. 11 (2).
Caws, P. 1977. *Sartre*. London.
Cohen, G. A. 1968. *Marx' Theory of History. A Defence*. Oxford.
Cooke, M. 1994. *Language and Reason*. Camb. Mass.
Dreyfuss, H. & P. Hoffmann 1981. "Sartre's changed Concept of Consciousness: From Lucidity to Opacity". In P. Schilpp (ed.), *The philosophy of Jean Paul Sartre*. La Salle, Ill.
Dux, G. 1982. *Die Logik der Weltbilder*. Frankfurt.
Elster, J. 1979. *Logic and Society*. Chichester.
Elster, J. 1986. *An Introduction to Karl Marx*. Cambridge.
Elster, J. 1989. *The Cement of Society*. Cambridge:
Engels, F. 1968. *Die Ursprung der Famile, des Privateigentums und des Staats.* Berlin.
Feenberg, A. 1991. *Critical Theory of Technology*. Oxford.
Feenberg, A. 1994a."The Technology Thesis Revisited: On The Critique of Power". *Inquiry*, vol. 37.
Feenberg, A. 1994b. *Marcuse or Habermas: Two Critiques of Technology*. (Working Paper no. 83, Center for Technology and Culture). Oslo.
Ferré, F. 1988. *The Philosophy of Technology*. Eng. Cliffs.
Føllesdal, D., Wallöe, L, Elster, J. 1986. *Argumentasjonsteori, språk og vitenskapsfilosofi*. Oslo.
Gadamer, H. G. 1973. "Hegels Dialektik des Selbstbewustseins". In H. F. Fulda & D. Henrich (eds.), *Materialen zu Hegels "Phänomenologie des Geistes"*. Frankfurt.
Gehlen, A. 1941. *Ein Bild von Menschen*. In K.-S. Rehberg (ed.), Gesamtausgabe, vol. 5, *Philosophische Anthropologie und Handlungslehre*. Frankfurt.
Gehlen, A. 1942. *Zur Systematik der Anthropologie*. In K. S. Rehberg (ed.), Gesamtausgabe, vol. 5, *Philosophische Anthropologie und Handlungslehre*. Frankfurt.
Gehlen, A. 1952a. *Das bild des Menschen im Lichte der Modernen Anthropologie*. In: K.-S.Rehberg (ed.), Gesamtausgabe, vol. 5, *Philosophische Anthropologie und Handlungslehre*. Frankfurt.
Gehlen, A. 1952b. *Über die Geburt der Freiheit aus der Entfremdung*. In: K.-S. Rehberg, (ed.), Gesamtausgabe, vol. 5, Philosophische Anthropologie und Handlungslehre.. Frankfurt.
Gehlen, A. 1968. *Ein Anthropologisches Modell*. In K.-S. Rehberg (ed.), Gesamtausgabe, vol. 5, *Philosophische Anthopologie und Handlungslehre*. Frankfurt.
Gehlen, A. 1977. *Urmensch und Spätkultur*. Frankfurt.

Gehlen, A. 1980. *Man in the Age of Technology (Die Seele im technischen Zeitalter).* New York.
Gehlen, A. 1988. Man. *His Nature and Place in the World (Der Mensch, seine Natur und Stellung in der Welt).* New York.
Gershuny, J. 1983. *Social Innovation and the Division of Labour.* Oxford.
Geuss, R. 1981. *The Idea of a Critical Theory.* Cambridge.
Gibson, J. J. 1966. *The senses considered as a perceptual system.* Boston.
Glaser, W. R. 1972. *Soziales und instrumentales Handeln.* Stuttgart.
Gould, S. J. 1978. *Ever since Darwin.* Harmondsworth.
Habermas, J. 1958. "Anthropologie". In *Das Fischer Lexicon Philosophie.* Frankfurt.
Habermas, J. 1970/71. "Vorlesungen zur einer sprachtheoretischen Grundlegung der Soziologie". In *Vorstudien und Ergänzungen zur Theorie des kommunikativen Handelns.* 1984. Frankfurt.
Habermas, J. 1971a. *Toward a Rational Society.* London.
Habermas, J. 1971b. Theorie und Praxis, 4 ed.. Frankfurt.
Habermas, J. 1972. "Wahrheitstheorien". *In Vorstudien und Ergänzungen zur Theorie des kommunikativen Handelns* 1984. Frankfurt.
Habermas, J. 1974. Theory and Practice. London.
Habermas, J. 1979. Communication and the Evolution of Society. London.
Habermas, J. 1981. "Nachgeahmte Substansialität". In: *Philosophisch-Politische Profile.* Frankfurt.
Habermas, J. 1982. "A Reply to my Critics". In J. Thompson & D Held (eds.), *Habermas. Critical Debates.* London.
Habermas, J. 1983. "Die Philosophie als Platzhalter und Interpret". In *Moralbewusstsein und kommunikatives Handeln.* Frankfurt.
Habermas, J. 1985 a. *Der philosophische Diskurs der Moderne.* Frankfurt.
Habermas, J. 1985 b. "Bemerkungen zur Beginn einer Vorlesung". In: *Die neue Unübersichtlichkeit.* Frankfurt.
Habermas, J. 1985c. "Questions and Counterquestions". In R. J. Bernestein (ed.), *Habermas and Modernity.* Cambridge.
Habermas, J. 1987a. *Knowledge and Human Interests.* Cambridge.
Habermas, J. 1987b. *The theory of Communicative Action.* Vol. II. London.
Habermas, J. 1988. *Nachmetaphysisches Denken.* Frankfurt.
Habermas, J. 1990. *Strukturwandel der Öffentlichkeit,* Frankfurt. Habermas, J. 1991. "A Reply". In A. Honneth & H. Joas (eds.), *Communicative Action.* Camb Mass.
Habermas, J. 1994. *Between Facts and Norms.* Camb. Mass.
Hallpike, C. R. 1986. *The Principles of Social Evolution.* Oxford.
Hartmann, K. 1966. *Sartres Sozialphilosophie.* Berlin.
Hartnack, J. 1979. *Fra Kant til Hegel.* København.

Harvey, D. 1989. *The Condition of Postmodernity*. Oxford.
Hegel, G. W. F. 1821. *Grundlinien der Philosophie des Rechts*. Berlin.
Hegel, G. W. F. 1932. *Jenaer Realphilosophie* I. (ed.) J. Hoffmeister. Hamburg:
Hegel, G. W. F. 1977. *Phenomenology of Spirit*. Oxford.
Heidegger, M. 1927. *Sein und Zeit*. Tübingen.
Heidegger, M. 1977. "The Age of the World Picture". In: *The Question Concerning Technology*. New York.
Herf, J. 1984. *Reactionary Modernism*. Camb. Mass.
Hesse, M. 1982. "Science and Objectivity". In: J. Thompson, & D. Held (eds.), *Habermas. Critical Debates*. London.
Honneth, A. & H. Joas (eds.) 1991. *Communicative Action*. Cambridge.
Honneth, A. 1991. *The Critique of Power*. Camb. Mass.
Horkheimer, M. 1931. "Die Gegenwärtige Lage der Sozialphilosophie und die Aufgaben eines Instituts für Sozialforschung". *Gesammelte Schriften*, vol. III. Frankfurt.
Horkheimer, M. 1937. "Traditionelle und kritische Theorie". *Zeitschrift für Sozialforschung*, vol. 6.
Horton, R. 1982. "Tradition and Modernity Revisited". In M. Hollis & S. Lukes (eds.), *Rationality and Relativism*. Oxford.
Hörning, K. 1988. "Technik im Alltag und die Widersprüche des Alltäglichen". In B. Joerges (ed.), *Technik im Alltag*. Frankfurt.
Ihde, D. 1990. *Technology and the Lifeworld*. Bloomington and Indianapolis.
Ihde, D. 1991. *Instrumental Realism*. Bloomington and Indianapolis.
Jameson, F. 1971. *Marxism and Form*. Princeton.
Joerges, B. 1988. "Gerätetechnik und Alltagshandeln". In: B. Joerges (ed.), *Technik im Alltag*. Frankfurt.
Kant, I. 1787. *Kritik der reinen Vernunft*. Riga.
Kapp, E. 1877. *Grundlinien einer Philosophie der Technik*. Braunschweig.
Knecht, I. 1975. *Die Theorie der Entfremdung bei Sartre und Marx*. Meisenheim.
Koyeve, A. 1958. *Hegel. Versuch einer Vergegenwärtigung seines Denkens*. Stuttgart.
Krogh, T. 1985 *Transcendental Affinitet*. Ph.D thesis, Oslo.
Krogh, T. 1991a. *Fra Frankfurt til Hollywood*. Oslo.
Krogh, T. 1991b. Types of Critique of Technology. In: C. Landström (ed.), *Intellectuals reading Technology: Science, technology, ideology, culture*. Gothenburg.
Lambert, K. & G. Brittan. 1987. *An Introduction to the Philosophy of Science*. Ascadero.
Landes, D. 1966. Introduction. In D. Landes (ed.), *The Rise of Capitalism*. New York.

Landes, D. 1970. *The Unbound Prometheus.* Cambridge.
Lenk, H. 1982. *Zur Sozialphilosophie der Technik.* Frankfurt.
Leroi-Gourhan, A. 1980. *Hand und Wort.* Frankfurt.
Linde, H. 1972. *Sachdominans in Sozialstrukturen.* Tübingen.
Locke, J. 1961. *An Essay Concerning Human Understanding.* London.
Luhmann, N. & J. Habermas 1971. *Theorie der Gesellschaft oder Sozialtechnologie - Was leistet die Systemforschung?* Frankfurt.
Luhmann, N. 1988. *Macht.* Stuttgart.
Lukacs, G. 1973. *Der Junge Hegel, vol I-II.* Frankfurt.
Lyons, J. 1988. "Origins of language". In A. C. Fabian (ed.), *Origins.* Cambridge.
Mantoux, P. 1961. *The Industrial Revolution in the Eighteenth Century (rev. ed.).* London.
Marcuse, H. 1941. "Some Social Implications of Modern Technology". Zeitschrift fur Sozialforschung vol. IX (3).
Marcuse, H. 1964. *One Dimensional Man.* Boston.
Marx, K. 1953. *Grundrisse zur Kritik der politischen ökonomie.* Berlin.
Marx, K. 1863. Letter to F. Engels 28/1 1863. Marx-Engels Werke vol. 30 Berlin.
Marx, K. 1964. *Das Elend der Philosophie.* Marx-Engels Werke vol. 4 Berlin.
Marx, K. 1965. *The Capital*, vol 1. Moscow.
Marx, K. 1966. Randglossen zur Adolph Wagners "Lehrbuch der politischen Ökonomie" (1873). Marx-Engels Werke vol 19 Berlin.
Marx, K. 1969a. *Die deutsche Ideologie.* Marx-Engels Werke vol 4. Berlin.
Marx, K. 1969b. *Resultate des unmittelbaren Produktionprozesses.* Frankfurt.
Marx, K. 1973. *Ökonomisch-Philosophische Manuskripte.* Marx-Engels Werke Egänzungsband. Erster Teil. Berlin.
McCarthy, T. 1978. *The Critical Theory of Jürgen Habermas.* Camb. Mass.
McCarthy, T. 1991. *Ideals and Illusions.* Cambridge, Mass.
McKendrick, N. 1966. "Josiah Wedgwood and factory discipline". In D. Landes (ed.), *The Rise of Capitalism.* New York.
Merleau-Ponty, M. 1962. *The Phenomenology of Perception.* London.
Merleau-Ponty, M. 1968. *Die Abenteuer der Dialektik.* Frankfurt.
Moser, I. B. 1993. *Teknologier i samfunnsteori: Forskyvninger og forflytninger.* Oslo.
Oakley, K. P. 1959. "Skill as a Human Posssesion". In G. Sabine et al. *A History of Technology.* Oxford.
Overend, T. 1978." Enquiry and Ideology: Habermas's Trichotomous Conception of Science". *Philosophy of Social Science,* vol 8.
Pagden, A. (ed.) 1987. *The Languages of Politics in early Modern Europe.* Cambridge.

Popitz, H., H. P. Bahrdt et al. 1957. *Technik und Industriearbeit*. Tübingen.
Popper, K. 1969. *Conjectures and Refutations*. London.
Pot, J. H. J. van der 1985. *Die Bewertung des technischen Fortschritts*. Assen/Mastricht.
Rammert, W. 1988. "Technisierung im Alltag". In B. Joerges (ed), *Technik im Alltag*. Frankfurt.
Reuleaux, F. 1875. *The Kinematics of Machinery*. London.
Ringer, F. 1969. *The Decline of the German Mandarins*. Cambr., Mass.
Rosenberg, N. 1982. "Marx as a student of technology". In *Inside the Black Box: Technology and Economics*. Cambridge.
Sartre, J.-P. 1963. *The Problem of Method*. London.
Sartre, J.-P. 1976. *Critique of Dialectical Reason*. London - New York.
Schelsky, H. 1961. *Der Mensch in der wissenschaftlichen Zivilisation*. Köln-Opladen.
Schmidt, A. 1963. *Der Begriff der Natur in der Lehre von Marx*. Frankfurt.
Schmidt, H. 1953. "Die Entwicklung der Technik als Phase der Wandlung des Menschen". *Zeitschrift des Vereins der deutschen Ingenieure* vol. 96 (5).
Schnädelbach, H. 1972. "Über den Realismus". *Zeitschrift für allgemeine Wissenschaftstheorie* vol. III/1.
Schumpeter, J. 1961. *Capitalism, socialism and democracy*. London.
Sejersted, F. 1990. *Er det mulig å styre utviklingen?* (Working Paper nr. 11. Center for Technology and Culture) Oslo.
Sellars, W. 1967. "Some Remarks on Kant's Theory of Experience". The Journal of Philosophy.
Sieferle, R. P. 1984. *Fortschrittsfeinde?* München.
Skirbekk, G. 1982. "Rationaler Konsens und ideale Sprechsituation als Geltungsgrund? Über Recht und Grenze eines tranzendentalpragmatisches Geltungskonzepts." In W. Kühlmann & D. Böhler (eds.), *Kommunikation und Reflexion*. Frankfurt.
Skirbekk, G. 1993. *Rationality and Modernity*. Oslo and Oxford.
Stockmann, N. 1978. "Habermas, Marcuse and the Aufhebung of Science and Technology". *Philosophy of Social Science*, vol. 8.
Taylor, C. 1975. *Hegel*. Cambridge.
Thao, T. D. 1973. *Investigations into the Theory of Language and Consciousness*. Dordrecht.
Ure, A. 1861. *The Philosophy of Manufactures*. London.
Usher, A. P. 1954. *A History of Mechanical Inventions* (rev. ed.). Cambridge.
Warnock, M. 1982."Historical Explanations in "The Critique of Dialectical Reason"". In G. H. R. Parkinson (ed.), *Marx and Marxisms*. Cambridge.

Weber, M. 1968. *The Theory of Social and Economic Organization*. York
Weber, M. 1985." Über einige Kategorien der Verstehende Soziologie". In J. Winkelmann (ed.), *Gesammelte Aufsätze zur Wissenschaftslehre*. Tübingen.
Wellmer, A. 1977. "Kommunikation und Emanzipation. Überlegungen zur "sprachanalytischen Wende" der kritischen Theorie". In U. Jaeggi & A. Honneth (eds.), *Theorien des historischen Materialismus*. Frankfurt.
White, L. 1963. *Medieval Technology and social Change*. Oxford.
White, S. K. 1988. *The recent work of Jürgen Habermas*. Cambridge.
Winkelmann, R. 1982. "Kommentar". In W. Reiner (ed.), Karl Marx: *Exzerpte über Arbeitsteilung, Maschinerie und Industrie*. Frankfurt.
Winner, L. 1977. *Autonomous Technology*. Camb. Mass.
Winner, L. 1986. *The Whale and the Reactor*. Chicago.
Østerberg, D. 1974. *Notater til materiellbegrepet*. Oslo.
Østerberg, D. 1993. *Jean-Paul Sartre. Filosofi, politikk, kunst, privatliv*. Oslo.

# Index

Adorno Th, 107-108; 110-111; 114; 145; 147-150
Althusser L, 65
anthropology, 9; 17; 22; 25-27; 32; 39; 85; 89-90; 104; 156
Arkwright R, 79
Armstrong-Kelley G, 4; 47
Ayers M, 51

Babbage C, 70; 72; 74
Bahrdt H.P, 9
Balibar E, 64-65
Baruzzi A, 19
Bauer B, 57
Beck B, 28; 32; 54; 156
Beckmann S, 12
Berg M, 76
Berger P, 25; 148; 167
Bijker W.E, 8
Blumenberg H, 4; 18-19; 41; 125; 184
Borgmann A, 140
Boulton M, 79
Braudel F, 87
Brittan G, 111
Bruland K, 75; 76; 77;

Carlyle T, 140
Cassirer E, 42
Caws P, 83
cognitive interest, 99; 101; 103; 104; 106-107; 111-113; 115; 120-121; 143; 146
Cohen G, 59
communicative action, 14; 99; 100; 101; 121; 128; 130-133; 136; 144; 146; 153-154; 162; 178-179; 185; 190

communicative theory, 140; 152-153; 155; 191; 193
Cooke M, 144; 164; 175; 180

de Beauvoir S, 95
Descartes R, 18; 41; 100; 144
Dewey J, 116
Dreyfuss H, 83
Durkheim E, 137; 166; 168
Dux G, 52

Ebeling H, 106
Ellul J, 11; 121
Elster J, 48; 59; 114; 133
experimental action, 119

Feenberg A, vii; 58; 186-189
Ferré F, 1-3
Feuerbach L, 57
Flaubert G, 85
form of mutual understanding, 190
formal subsumption, 67; 68
Frankfurt School, 108; 111-112; 135; 144; 150; 178
Freyer H, 121-122
Føllesdal D, 114

Gadamer H-G, 4; 47; 151
Galileo G, 108
Gehlen A, 2; 10-11; 17-20; 23; 25-36; 39; 53; 113; 139; 154; 155;
Genet J, 84
Gershuny J, 13
Geuss R, 111
Gibson J, 51

Habermas J, vii; 2; 13; 17; 25; 35; 39; 44; 88; 93; 99-101; 103-136;

140; 143-152; 154-167; 169; 171-186; 188-193
Hartmann K, 84; 86
Heidegger M, 10-11; 18; 22; 55; 95; 108; 181
Henrich D, 106
Herder J.G, 27; 43
Herf J, 10
Hesse M, 119-120
Hoffmann P, 83
Honneth A, 101; 123; 128; 130; 133-136; 159; 162; 165; 169; 189
Horkheimer M, 54; 114; 143; 145; 147-148; 150
Hughes T, 8
Husserl E, 22; 100; 144; 148-149; 151; 184
Hutton C, 75

ideology, 10; 13; 96; 99; 101; 104; 121; 123; 126-128; 135; 150; 184
Ihde D, 10
inert, 81-82; 84-86; 88; 90-91; 94-96
instrumental action, 13; 17; 103; 105-106; 112-116; 118-120; 131-132; 153; 156

Jameson F, 82-84
Joas H, 162; 164;
Joerges B, 6;

Kant I, 19; 39; 41; 42; 45; 47; 49; 86; 89; 100; 105; 106; 107; 133; 144; 147; 149; 151; 153; ;
Kapp E, 9; 11; 17; 18; 19; 20; 21; 22; 23; 29; 30; 31; 34; 35
Knecht I, 47; 84; 87
Krogh T, vii; 11; 41; 42; 52; 113
Kuhn T, 120

labour, 1; 3; 8; 12-13; 30; 33; 39; 42-46; 48-54; 56-57; 59-72; 74-82; 88-91; 94; 99; 101; 105-106; 108; 111; 114-115; 121; 123; 128; 130-134; 136-137; 143; 145-147; 150-151; 153-157; 169; 175; 177; 179; 181; 188; 190-191
Lambert K, 111
Landes D, 78; 79
Lenk K, 12; 18-19; 25
lifeworld, 6; 10; 13; 22; 31-32; 69; 81; 126-130; 134-135; 151; 154-155; 157-158; 160-169; 171-186; 189-194
Locke J, 52
Lorenz K, 28
Luckmann T, 148; 167
Luhmann N, 2; 39; 159; 183-184; 188
Lukacs G, 10; 42-43; 112; 140; 147; 150; 172; 190

machine, 3; 6; 24; 34; 52; 58; 66; 68-77; 80; 94-96; 115; 121; 181; 184
MacKenzie D, 59
Mantoux P, 80
Marcus G, 147
Marcuse H, 11; 112-113; 135; 187
Marx K, 2-3; 5-6; 12-13; 19; 26; 39; 42-46; 52; 55; 57-59; 61-80; 82; 86; 88; 92; 100; 108; 116; 123-124; 140; 146-147; 149-151; 153-154; 158-159; 171; 183
master and slave, 47
materiality, vii; 5-6; 9; 12-13; 39; 42; 53; 55-56-57-76-77; 80-82; 84; 87-88; 91-92; 94; 96; 100; 112; 128; 136; 140; 153; 155; 167; 177-179; 181-182; 185-186; 188; 191-194
matter, 18; 25; 29; 35; 51; 55-56; 63; 81-82; 84-87; 91; 94-95; 107; 111; 115; 131-132; 134; 165; 188
McKendrick N, 77; 78; 79
Mead G.H, 144; 149; 168
media, 14; 43; 70; 81; 141; 165; 168; 173-179; 182-189; 191-192
medium, 43-44; 56; 84; 88; 93; 137; 140; 166; 179; 184; 186-189; 193
methodological individualism, 148-149; 159; 160

money, 14; 60-61; 92-93; 173-174; 177; 183-187; 189; 192
Moses R, 188
Mumford N, 11; 91
Myrdal G, 192

Oakley K.P, 54
Origen, 21
Overend T, 103

Parsons C, 93; 130; 173; 188
philosophy of consciousness, 105; 143; 144; 145; 148
Popitz H, 9
Popper K., 107-110
positivism debate, 107
Pot P van der, 21
power, 6; 9; 14; 34; 60-61; 65; 73; 75-76; 78-80; 83; 91; 95; 134-135; 161; 168; 173-175; 177; 179; 183-187; 189; 192; 194
pragmatic turn, 144; 145
projection, 20-23; 29-30; 34-35
Protagoras, 21; 22

Rammert W, 13
real subsumption, 66-68; 71; 73; 76
relief, 23; 29-31; 33; 35; 39; 113; 122; 173-174; 184-186; 188; 192
Reuleaux F, 74-75
Ricardo D, 60
Ringer F, 9

Sartre J-P, 2-3; 5; 12; 39; 55; 64; 81-91; 93-96; 100; 144; 148; 154; 167-177
scarcity, 84; 85; 89; 90; 91; 94
Scheler M, 10; 26-27
Schelsky H, 121-122
Schilpp A,
Schmidt H, 3; 33-34; 64
Schütz A, 148; 167
SCOT programme, 8-9; 13
sensible intuition, 41-42; 51
Sieferle R.P, 11; 69

Skirbekk G, vii; 104-105; 116-117; 162-164
Smith A, 45; 60; 71; 159
social integration, 154; 158; 164-165; 168-169; 174-175; 178-179; 182-185; 189-190; 192
Sommer M, 107
Spencer H, 21
St. Victor H, 21
system, 6; 12-13; 21; 39; 43-45; 47; 57-58; 64-66; 68-70; 73; 75-76; 78-80; 110; 114-115; 122; 124; 128-130; 133; 150; 154-155; 157-166; 168-169; 171-176; 182-185; 187; 189-192
system integration, 154-155; 158; 165; 168; 171; 174-175; 183-185

Taylor C, 43-44; 46; 74
technocracy, 35; 101; 121-126; 130; 132; 134-136; 187
*Technologiephilosophie*, 9-10; 20
Thao T. D, 156-157
tool, 21-22; 33; 39; 44; 52; 56; 71-73; 75; 91; 95; 161; 181
type of action, 115; 129; 132; 140

uncoupling, 30; 161; 165-166; 168; 172; 182
Ure A, 70; 79-80
Usher A. P, 72; 74-76

Veblen T, 121

Walløe L, 114
Warnock M, 85
Weber M, 112; 122-123; 129-130; 132-133; 136-140; 149; 150
Wedgwood J, 78-79
Wellmer A, 150; 151
White L, 69
White S. K, 144
Winkelmann R, 76
Winner L, 10-11; 121; 188
Østerberg D, vii; 5; 83; 136